Alcoholic Fuels

CHEMICAL INDUSTRIES

A Series of Reference Books and Textbooks

Consulting Editor

HEINZ HEINEMANN
Berkeley, California

1. *Fluid Catalytic Cracking with Zeolite Catalysts,* Paul B. Venuto and E. Thomas Habib, Jr.
2. *Ethylene: Keystone to the Petrochemical Industry,* Ludwig Kniel, Olaf Winter, and Karl Stork
3. *The Chemistry and Technology of Petroleum,* James G. Speight
4. *The Desulfurization of Heavy Oils and Residua,* James G. Speight
5. *Catalysis of Organic Reactions,* edited by William R. Moser
6. *Acetylene-Based Chemicals from Coal and Other Natural Resources,* Robert J. Tedeschi
7. *Chemically Resistant Masonry,* Walter Lee Sheppard, Jr.
8. *Compressors and Expanders: Selection and Application for the Process Industry,* Heinz P. Bloch, Joseph A. Cameron, Frank M. Danowski, Jr., Ralph James, Jr., Judson S. Swearingen, and Marilyn E. Weightman
9. *Metering Pumps: Selection and Application,* James P. Poynton
10. *Hydrocarbons from Methanol,* Clarence D. Chang
11. *Form Flotation: Theory and Applications,* Ann N. Clarke and David J. Wilson
12. *The Chemistry and Technology of Coal,* James G. Speight
13. *Pneumatic and Hydraulic Conveying of Solids,* O. A. Williams
14. *Catalyst Manufacture: Laboratory and Commercial Preparations,* Alvin B. Stiles

Alcoholic Fuels

Shelley Minteer

Saint Louis University
Missouri

CRC Press
Taylor & Francis Group
Boca Raton London New York

CRC Press is an imprint of the
Taylor & Francis Group, an **informa** business
A TAYLOR & FRANCIS BOOK

First published 2006 by Taylor & Francis

Published 2019 by CRC Press
Taylor & Francis Group
6000 Broken Sound Parkway NW, Suite 300
Boca Raton, FL 33487-2742

© 2006 by Taylor & Francis Group, LLC
CRC Press is an imprint of Taylor & Francis Group, an Informa business

First issued in paperback 2019

No claim to original U.S. Government works

ISBN 13: 978-0-367-45357-2 (pbk)
ISBN 13: 978-0-8493-3944-8 (hbk)

Visit the Taylor & Francis Web site at
http://www.taylorandfrancis.com

and the CRC Press Web site at
http://www.crcpress.com

Library of Congress Cataloging-in-Publication Data

Alcoholic fuels / edited by Shelley Minteer.
     p. cm.
  ISBN 0-8493-3944-8 (alk. paper)
  1. Alcohol as fuel. I. Minteer, Shelley D. II. Title.

TP358.A4445 2006
662'.6692--dc22
                                      2005056058

Library of Congress Card Number 2005056058

Preface

In the 1880s, Henry Ford developed a prototype automobile (the quadracycle) that could be operated with ethanol as fuel. Historians say that Ford always believed that the Model T and his future cars would use alcohol as fuel because it was a renewable energy source and would boost the agricultural economy. Over a century later, research has finally brought us to the point at which using alcohol-based fuels for transportation applications is a reality. Over the last two decades, research on alcoholic fuels as alternative and renewable energy sources has exponentially increased. Some of these alcoholic fuels (e.g., methanol and ethanol) have been introduced into the market as alcohol-gasoline blends for combustion engines, but research has also focused on employing these alcohols as fuels for alternative energy platforms, such as fuel cells. This book will provide a comprehensive text to discuss both the production of alcoholic fuels from various sources and the variety of applications of these fuels, from combustion engines to fuel cells to miniature power plants (generators) for farms.

Currently, there is no text on alcoholic fuels. The books on the market that come close are *Biomass Renewable Energy, Fuels, and Chemicals* (1998) and *Renewable Energy: Sources for Fuels and Electricity* (1992). Neither of these texts focuses on alcoholic fuels. Both books focus on the production of all renewable energy sources and have sections on the production of alcoholic fuels, but they do not include the necessary information to see the history and future of alcoholic fuels from both production and application viewpoints. This book is comprised of edited chapters from experts and innovators in the field of alcohol fuels. The book is broken down into three sections. The first section focuses on the production of methanol, ethanol, and butanol from various biomasses including corn, wood, and landfill waste. The second section focuses on blended fuels. These are the fuels that mix alcohols with existing petroleum products, such as gasoline and diesel. The final section focuses on applications of alcoholic fuels. This includes different types of fuel cells, reformers, and generators. The book concludes with a chapter on the future of alcohol-based fuels. The book is intended for anyone wanting a comprehensive understanding of alcohol fuels. Each chapter has sufficient detail and provides scientific references sufficient for researchers to get a detailed perspective on both the production of alcoholic fuels and the applications of alcoholic fuels, but the chapters themselves are comprehensive in order to provide the reader with an understanding of the history of the technology and how each application plays an important role in removing our dependency on oil and environmentally toxic power sources, such as batteries. The book is intended to be a supplementary text for graduate courses on alternative energy, power sources, or fuel cells. There are books on each of these subjects, but no book that ties them together. To really understand alcohol-based fuel cells, you need a thorough understanding of how the alcohol is produced and purified. On the other hand, a scientist whose focus is on improving the production of ethanol needs to have a thorough understanding of how the alcohol is being used.

Editor

Shelley Minteer received her Ph.D. in chemistry in 2000 from the University of Iowa. She has been on the faculty of the Department of Chemistry at Saint Louis University since 2000 and was promoted to the rank of associate professor in 2005. She also holds a second appointment in the Department of Biomedical Engineering. Since arriving at Saint Louis University, Dr. Minteer's research has focused on the development of efficient alternative energy sources, specifically alcohol/oxygen biofuel cells.

Contributors

Nick L. Akers
Akermin, Incorporated
St. Louis, Missouri

Hans P. Blaschek
Biotechnology & Bioengineering
  Group
Department of Food Science &
  Human Nutrition
University of Illinois
Urbana, Illinois

Rodney J. Bothast
National Corn-to-Ethanol
  Research Center
Southern Illinois University-
  Edwardsville
Edwardsville, Illinois

Hachull Chung
Department of Chemistry
University of Iowa
Iowa City, Iowa

Michael A. Cotta
Fermentation Biotechnology
  Research Unit
National Center for Agricultural
  Utilization Research,
  Agricultural Research Service
U.S. Department of Agriculture
Peoria, Illinois

Gregory W. Davis, Ph.D. P.E.
Advanced Engine Research
  Laboratory and Department of
  Mechanical Engineering
Kettering University
Flint, Michigan

Pilar Ramírez de la Piscina
Inorganic Chemistry Department
Universitat de Barcelona
Barcelona, Spain

Bruce S. Dien
Fermentation Biotechnology
  Research Unit
National Center for Agricultural
  Utilization Research,
  Agricultural Research Service
U.S. Department of Agriculture
Peoria, Illinois

Fatih Dogan
Department of Materials Science and
  Engineering
University of Missouri-Rolla
Rolla, Missouri

Drew C. Dunwoody
Department of Chemistry
University of Iowa
Iowa City, Iowa

Thaddeus C. Ezeji
Biotechnology & Bioengineering
  Group
Department of Food Science &
  Human Nutrition
University of Illinois
Urbana, Illinois

André P.C. Faaij
Utrecht University/Copernicus
  Institute of Sustainable Development
  and Innovation
Utrecht, The Netherlands

Robert Haber
One Accord Food Pantry, Inc.
Troy, New York

Dr. Carlo N. Hamelinck
Ecofys
Utrecht, The Netherlands

Luke Haverhals
Department of Chemistry
University of Iowa
Iowa City, Iowa

Narcís Homs
Inorganic Chemistry Department
Universitat de Barcelona
Barcelona, Spain

Hans-Joachim G. Jung, Ph.D.
U.S. Department of Agriculture
Agricultural Research Service-
  Plant Science Research
Department of Agronomy/Plant
  Genetics
University of Minnesota
St. Paul, Minnesota

Patrick Karcher
Biotechnology & Bioengineering
  Group
Department of Food Science &
  Human Nutrition
University of Illinois
Urbana, Illinois

JoAnn F. S. Lamb
U.S. Department of Agriculture
Agricultural Research Service-
  Plant Science Research
Department of Agronomy/Plant
  Genetics
University of Minnesota
St. Paul, Minnesota

Johna Leddy
Department of Chemistry
University of Iowa
Iowa City, Iowa

Nancy N. Nichols
Fermentation Biotechnology
  Research Unit
National Center for Agricultural
  Utilization Research,
  Agricultural Research Service
U.S. Department of Agriculture
Peoria, Illinois

Nasib Qureshi
U.S. Department of Agriculture
National Center for Agricultural
  Utilization Research,
  Fermentation/Biotechnology
Peoria, Illinois

Deborah A. Samac
U.S. Department of Agriculture
Agricultural Research Service-
  Plant Science Research
Department of Plant Pathology
University of Minnesota
St. Paul, Minnesota

Sabina Topcagic
Department of Chemistry
Saint Louis University
St. Louis, Missouri

Becky L. Treu
Department of Chemistry
Saint Louis University
St. Louis, Missouri

William H. Wisbrock, President
Biofuels of Missouri, Inc.
St. Louis, Missouri

Table of Contents

SECTION II   Blended Fuels

SECTION III   Applications of Alcoholic Fuels

Chapter 15
Nick L. Akers

1 Alcoholic Fuels: An Overview

Shelley D. Minteer
Saint Louis University, Missouri

CONTENTS

Abstract Alcohol-based fuels have been used as replacements for gasoline in combustion engines and for fuel cells. The four alcohols that are typically used as fuels are methanol, ethanol, propanol, and butanol. Ethanol is the most widely used fuel due to its lower toxicity properties and wide abundance, but this chapter introduces the reader to all four types of fuels and compares them.

INTRODUCTION

Alcohol-based fuels have been important energy sources since the 1800s. As early as 1894, France and Germany were using ethanol in internal combustion engines. Henry Ford was quoted in 1925 as saying that ethanol was the fuel of the future [1]. He was not the only supporter of ethanol in the early 20th century. Alexander Graham Bell was a promoter of ethanol, because the decreased emission to burning ethanol [2]. Thomas Edison also backed the idea of industrial uses for farm products and supported Henry Ford's campaign for ethanol [3]. Over the years and across the world, alcohol-based fuels have seen short-term increases in use depending on the current strategic or economic situation at that time in the country of interest. For instance, the United States saw a resurgence in ethanol fuel during the oil crisis of the 1970s [4]. Alcohols have been used as fuels in three main ways: as a fuel for a combustion engine (replacing gasoline), as a fuel additive to achieve octane boosting (or antiknock) effects similar to the petroleum-based additives and metallic additives like tetraethyllead, and as a fuel for direct conversion of chemical energy into electrical energy in a fuel cell.

Alcohols are of the oxygenate family. They are hydrocarbons with hydroxyl functional groups. The oxygen of the hydroxyl group contributes to combustion. The four most simplistic alcoholic fuels are methanol, ethanol, propanol, and butanol. More complex alcohols can be used as fuels; however, they have not shown to be commercially viable. Alcohol fuels are currently used both in combustion engines and fuel cells, but the chemistry occurring in both systems is the same. In theory, alcohol fuels in engines and fuel cells are oxidized to form carbon dioxide and water. In reality, incomplete oxidation is an issue and causes many toxic by-products including carbon monoxide, aldehydes, carboxylates, and even ketones. The generic reaction for complete alcohol oxidation in either a combustion engines or a fuel cell is

$$C_xH_{2x+2}O + (\frac{3x}{2})O_2 \rightarrow xCO_2 + (x+1)H_2O$$

It is important to note this reaction occurs in a single chamber in a combustion engine to convert chemical energy to mechanical energy and heat, while in a fuel cell, this reaction occurs in two separate chambers (an anode chamber where the alcohol is oxidized to carbon dioxide and a cathode chamber where oxygen is reduced to water.)

METHANOL

Methanol (also called methyl alcohol) is the simplest of alcohols. Its chemical structure is CH_3OH. It is produced most frequently from wood and wood by-products, which is why it is frequently called wood alcohol. It is a colorless liquid that is quite toxic. The LD_{50} for oral consumption by a rate is 5628 mg/kg. The LD_{50} for absorption by the skin of a rabbit is 20 g/kg. The Occupational Safety and Health Administration (OSHA) approved exposure limit is 200 ppm for 10 hours. Methanol has a melting point of −98°C and a boiling point of 65°C. It has a density of 0.791 g/ml and is completely soluble in water, which is one of the hazards of methanol. It easily combines with water to form a solution with minimal smell that still has all of the toxicity issues of methanol. Acute methanol intoxication in humans leads to severe muscle pain and visual degeneration that can lead to blindness. This has been a major issue when considering methanol as a fuel. Dry methanol is also very corrosive to some metal alloys, so care is required to ensure that engines and fuel cells have components that are not corroded by methanol. Today, most research on methanol as a fuel is centered on direct methanol fuel cells (DMFCs) for portable power applications (replacements for rechargeable batteries), but extensive early research has been done on methanol–gasoline blends for combustion engines.

ETHANOL

Ethanol (also known as ethyl alcohol) is the most common of alcohols. It is the form of alcohol that is in alcoholic beverages and is easily produced from corn, sugar, or fruits through fermentation of carbohydrates. Its chemical structure is CH_3CH_2OH. It is less toxic than methanol. The LD_{50} for oral consumption by a rat is 7060 mg/kg [5]. The LD_{50} for inhalation by a rat is 20,000 ppm for 10 hours [6]. The NIOSH recommended exposure limit is 1000 ppm for 10 hours [7]. Ethanol is available in a pure form and a denatured form. Denatured ethanol contains a small concentration of poisonous substance (frequently methanol) to prevent people from drinking it. Ethanol is a colorless liquid with a melting point of −144°C and a boiling point of 78°C. It is less dense than water with a density of 0.789 g/ml and soluble at all concentrations in water. Ethanol is frequently used to form blended gasoline fuels in concentrations between 10–85%. More recently, it has been investigated as a fuel for direct ethanol fuel cells (DEFC) and biofuel cells. Ethanol was deemed the "fuel of the future" by Henry Ford and has continued to be the most popular alcoholic fuel for several reasons: (1) it is produced from renewable agricultural products (corn, sugar, molasses, etc.) rather than nonrenewable petroleum products, (2) it is less toxic than the other alcohol fuels, and (3) the incomplete oxidation by-products of ethanol oxidation (acetic acid (vinegar) and acetaldehyde) are less toxic than the incomplete oxidation by-products of other alcohol oxidation.

BUTANOL

Butanol is the most complex of the alcohol-based fuels. It is a four-carbon alcohol with a structure of $CH_3CH_2CH_2CH_2OH$. Butanol is more toxic than either methanol or ethanol. The LD_{50} for oral consumption of butanol by a rat is 790 mg/kg. The LD_{50} for skin adsorption of butanol by a rabbit is 3400 mg/kg. The boiling point of butanol is 118°C and the melting point is −89°C. The density of butanol is 0.81 g/mL, so it is more dense than the other two alcohols, but less dense than water. Butanol is commonly used as a solvent, but is also a candidate for use as a fuel. Butanol can be made from either petroleum or fermentation of agricultural products. Originally, butanol was manufactured from agricultural products in a fermentation process referred to as ABE, because it produced Acetone-Butanol and Ethanol. Currently, most butanol is produced from petroleum, which causes butanol to cost more than ethanol, even though it has some favorable physical properties compared to ethanol. It has a higher energy content than ethanol. The vapor pressure of butanol is 0.33 psi, which is almost an order of magnitude less than ethanol (2.0 psi) and less than both methanol (4.6 psi) and gasoline (4.5 psi). This decrease in vapor pressure means that there are less problems with evaporation of butanol than the other fuels, which makes it safer and more environmentally friendly than the other fuels. Butanol has been proposed as a replacement for ethanol in blended fuels, but it is currently more costly than ethanol. Butanol has also been proposed for use in a direct butanol fuel cell, but the efficiency of the fuel cell is poor because incomplete oxidation products easily passivate the platinum catalyst in a traditional fuel cell.

PROPANOL

Although propanols are three carbon alcohols with the general formula C_3H_8O, they are rarely used as fuels. Isopropanol (also called rubbing alcohol) is frequently used as a disinfectant and considered to be a better disinfectant than ethanol, but it is rarely used as a fuel. It is a colorless liquid like the other alcohols and is flammable. It has a pungent odor that is noticeable at concentrations as low as 3 ppm. Isopropanol is also used as an industrial solvent and as a gasoline additive for dealing with problems of water or ice in fuel lines. It has a freezing point of $-89°C$ and a boiling point of $83°C$. Isopropanol is typically produced from propene from decomposed petroleum, but can also be produced from fermentation of sugars. Isopropanol is commonly used for chemical synthesis or as a solvent, so almost 2M tons are produced worldwide.

CONCLUSIONS

In today's fuel market, methanol and ethanol are the only commercially viable fuels. Both methanol and ethanol have been blended with gasoline, but ethanol is the current choice for gasoline blends. Methanol has found its place in the market as an additive for biodiesel and as a fuel for direct methanol fuel cells, which are being studied as an alternative for rechargeable batteries in small electronic devices. Currently, butanol is too expensive to compete with ethanol in the blended fuel market, but researchers are working on methods to decrease cost and efficiency of production to allow for butanol blends, because the vapor pressure difference has environmental advantages. Governmental initiatives should ensure an increased use of alcohol-based fuels in automobiles and other energy conversion devices.

REFERENCES

1. Ford Predicts Fuel From Vegetation, *The New York Times*, Sept. 20, 1925, p. 24.
2. *National Geographic*, 31, 131, 1917.
3. Borth, C., *Chemists and Their Work*, Bobbs-Merrill, New York, 1928.
4. Kovarik, B., Henry Ford, Charles F. Kettering and the Fuel of the Future, *Automot. Hist. Rev.*, 32, 7–27, 1998.
5. *Toxicology and Applied Pharmacology*, Academic Press, Inc., 16, 718, 1970.
6. *Raw Material Data Handbook, Vol. 1: Organic Solvents*, Nat. Assoc. Print. Ink Res. Inst., 1, 44, 1974.
7. *National Institute for Occupational Safety and Health, U.S. Dept. of Health, Education, and Welfare, Reports and Memoranda*, DHHS, 92–100, 1992.

Section I

Production of Alcohol Fuels

2 Production of Methanol from Biomass*

Carlo N. Hamelinck

(currently working with Ecofys b.v. Utrecht,
The Netherlands)

André P.C. Faaij

(Utrecht University, Copernicus Institute of Sustainable
Development and Innovation, Utrecht, The Netherlands)

CONTENTS

* This chapter is broadly based on Hamelinck, C.N. and Faaij, A.P.C., Future prospects for production of methanol and hydrogen from biomass, *Journal of Power Sources*, 111, 1, 1–22, 2002.

INTRODUCTION

Methanol (CH_3OH), also known as methyl alcohol or wood alcohol, is the simplest alcohol. It can be used as a fuel, either as a blend with gasoline in internal combustion engines* or in fuel cell vehicles.** Also, methanol has a versatile function in the chemical industry as the starting material for many chemicals.

Methanol is produced naturally in the anaerobic metabolism of many varieties of bacteria and in some vegetation. Pure methanol was first isolated in 1661 by Robert Boyle by distillation of boxwood. In 1834, the French chemists Dumas and Peligot determined its elemental composition. In 1922, BASF developed a process to convert synthesis gas (a mixture of carbon monoxide and hydrogen) into methanol. This process used a zinc oxide/chromium oxide catalyst and required extremely vigorous conditions: pressures ranging from 300–1000 bar, and temperatures of about 400°C. Modern methanol production has been made more efficient through the use of catalysts capable of operating at lower pressures. Also the synthesis gas is at present mostly produced from natural gas rather than from coal.

In 2005, the global methanol production capacity was about 40 Mtonne/year, the actual production or demand was about 32 Mtonne (Methanol Institute 2005). Since the early 1980s, larger plants using new efficient low-pressure technologies are replacing less efficient small facilities. In 1984, more than three quarters of

---

* In Europe methanol may be blended in regular gasoline up to 5% by volume without notice to the consumer. Higher blends are possible like M85 (85% methanol with 15% gasoline) but would require adaptations in cars or specially developed cars. Moreover, blends higher than 5% require adaptations in the distribution of fuels to gas stations and at the gas stations themselves. Pure methanol is sometimes used as racing fuel, such as in the Indianapolis 500.

** Methanol can be the source for hydrogen via on board reforming. Direct methanol fuel cells are under development that can directly process methanol (van den Hoed 2004).

world methanol capacity was located in the traditional markets of North America, Europe, and Japan, with less than 10 percent located in "distant-from–market" developing regions such as Saudi Arabia. But from that time most new methanol plants have been erected in developing regions while higher cost facilities in more developed regions were being shut down. The current standard capacities of conventional plants range between 2000 and 3000 tonnes of methanol per day. However, the newest plants tend to be much larger, with single trains of 5000 tonnes/day in Point Lisas, Trinidad (start-up in 2004), 5000 tonnes/day in Dayyer, Iran (start-up in 2006), and 5000 tonnes/day in Labuan, Malaysia (start construction in 2006).

Methanol produced from biomass and employed in the automotive sector can address several of the problems associated with the current use of mineral oil derived fuels, such as energy security and greenhouse gas emissions.

This chapter discusses the technology for the production of methanol from biomass. For a selection of concepts, efficiencies and production costs have been calculated.

TECHNOLOGY

OVERVIEW

Methanol is produced by a catalytic reaction of carbon monoxide (CO), carbon dioxide (CO_2), and hydrogen (H_2). These gases, together called *synthesis gas*, are generally produced from natural gas. One can also produce synthesis gas from other organic substances, such as biomass. A train of processes to convert biomass to required gas specifications precedes the methanol reactor. These processes include pretreatment, gasification, gas cleaning, gas conditioning, and methanol synthesis, as are depicted in Figure 2.1 and discussed in Sections 2.2–2.6.

PRETREATMENT

Chipping or comminution is generally the first step in biomass preparation. The fuel size necessary for fluidized bed gasification is between 0 and 50 mm (Pierik et al. 1995). Total energy requirements for chipping woody biomass are approximately 100 kJ$_e$/kg of wet biomass (Katofsky 1993) down to 240 kW$_e$ for 25–50 tonne/h to 3 × 3 cm in a hammermill, which gives 17–35 kJ$_e$/kg wet biomass (Pierik et al. 1995).

The fuel should be dried to 10–15% depending on the type of gasifier. This consumes roughly 10% of the energy content of the feedstock. Drying can be

FIGURE 2.1 Key components in the conversion of biomass to methanol.

done by means of hot flue gas (in a rotary drum dryer) or steam (direct/indirect), a choice that among others depends on other steam demands within the process and the extent of electricity coproduction. Flue gas drying gives a higher flexibility toward gasification of a large variety of fuels. In the case of electricity generation from biomass, the integration in the total system is simpler than that of steam drying, resulting in lower total investment costs. The net electrical system efficiency can be somewhat higher (van Ree et al. 1995). On the other hand, flue gas drying holds the risk of spontaneous combustion and corrosion (Consonni et al. 1994). For methanol production, steam is required throughout the entire process, thus requiring an elaborate steam cycle anyway. It is not *a priori* clear whether flue gas or steam drying is a better option in methanol production. A flue gas dryer for drying from 50% moisture content to 15% or 10% would have a specific energy use of 2.4–3.0 MJ/ton water evaporated (twe) and a specific electricity consumption of 40–100 kWh$_e$/twe (Pierik et al. 1995). A steam dryer consumes 12 bar, 200°C (process) steam; the specific heat consumption is 2.8 MJ/twe. Electricity use is 40 kWh$_e$/twe (Pierik et al. 1995).

GASIFICATION

Through gasification solid biomass is converted into synthesis gas. The fundamentals have extensively been described by, among others, Katofsky (1993). Basically, biomass is converted to a mixture of CO, CO_2, H_2O, H_2, and light hydrocarbons, the mutual ratios depending on the type of biomass, the gasifier type, temperature and pressure, and the use of air, oxygen, and steam.

Many gasification methods are available for synthesis gas production. Based on throughput, cost, complexity, and efficiency issues, only circulated fluidized bed gasifiers are suitable for large-scale synthesis gas production. Direct gasification with air results in nitrogen dilution, which in turn strongly increases downstream equipment size. This eliminates the TPS (Termiska Processer AB) and Enviropower gasifiers, which are both direct air blown. The MTCI (Manufacturing and Technology Conversion International, affiliate of Thermochem, Inc.) gasifier is indirectly fired, but produces a very wet gas and the net carbon conversion is low. Two gasifiers are selected for the present analysis: the IGT (Institute of Gas Technology) pressurized direct oxygen fired gasifier and the BCL (Battelle Columbus) atmospheric indirectly fired gasifier. The IGT gasifier can also be operated in a *maximum hydrogen* mode by increasing the steam input. Both gasifiers produce medium calorific gas, undiluted by atmospheric nitrogen, and represent a very broad range for the H_2:CO ratio of the raw synthesis gas.

IGT Gasifier

The IGT gasifier (Figure 2.2) is directly heated, which implies that some char and/or biomass are burned to provide the necessary heat for gasification. Direct heating is also the basic principle applied in pressurised reactors for gasifying coal. The higher reactivity of biomass compared to coal permits the use of air instead of pure oxygen.

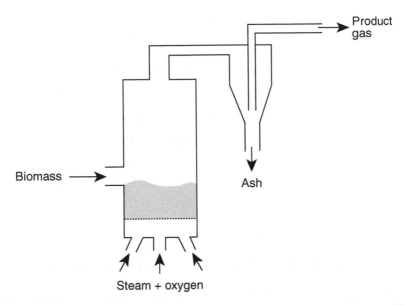

FIGURE 2.2 The directly heated, bubbling fluidized bed gasifier of IGT (Katofsky 1993).

This could be fortuitous at modest scales because oxygen is relatively costly (Consonni and Larson 1994a). However, for the production of methanol from biomass, the use of air increases the volume of inert (N_2) gas that would have to be carried through all the downstream reactors. Therefore, the use of oxygen thus improves the economics of synthesis gas processing. Air-fired, directly heated gasifiers are considered not to be suitable before methanol production.

This gasifier produces a CO_2 rich gas. The CH_4 fraction could be reformed to hydrogen, or be used in a gas turbine. The H_2:CO ratio (1.4:1) is attractive to produce methanol, although the large CO_2 content lowers the overall yield of methanol. The pressurized gasification allows a large throughput per reactor volume and diminishes the need for pressurization downstream, so less overall power is needed.

The bed is in a fluidized state by injection of steam and oxygen from below, allowing a high degree of mixing. Near the oxidant entrance is a combustion zone with a higher operation temperature, but gasification reactions take place over the whole bed, and the temperature in the bed is relatively uniform (800–1000 °C). The gas exits essentially at bed temperature. Ash, unreacted char, and particulates are entrained within the product gas and are largely removed using a cyclone.

An important characteristic of the IGT synthesis gas is the relatively large CO_2 and CH_4 fractions. The high methane content is a result of the nonequilibrium nature of biomass gasification and of pressurized operation. Relatively large amounts of CO_2 are produced by the direct heating, high pressure, and the high overall O:C ratio (2:1). With conventional gas processing technology, a large CO_2 content would mean that overall yields of fluid fuels would be relatively low. The synthesis gas has an attractive H_2:CO ratio for methanol production, which

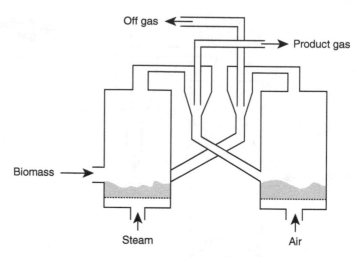

FIGURE 2.3 The indirectly heated, twin-bed gasifier of BCL (Katofsky 1993).

reduces the need for a shift reactor. Since gasification takes place under pressure, less downstream compression is needed.

When operated with higher steam input the IGT gasifier produces a product gas with a higher hydrogen content. This *maximum hydrogen* mode is especially useful if hydrogen would be the desired product, but the H_2:CO ratio is also better for methanol production. However, the gasifier efficiency is lower and much more steam is needed.

BCL Gasifier

The BCL gasifier is indirectly heated by a heat transfer mechanism as shown in Figure 2.3. Ash, char, and sand are entrained in the product gas, separated using a cyclone, and sent to a second bed where the char or additional biomass is burned in air to reheat the sand. The heat is transferred between the two beds by circulating the hot sand back to the gasification bed. This allows one to provide heat by burning some of the feed, but without the need to use oxygen, because combustion and gasification occur in separate vessels.

Because of the atmospheric pressure, the BCL gasifier produces a gas with a low CO_2 content, but consequently containing a greater number of heavier hydrocarbons. Therefore, tar cracking and reforming are logical subsequent steps in order to maximize CO and H_2 production. The reactor is fast fluidized allowing throughputs equal to the bubbling fluidized IGT, despite the atmospheric opera-tion. The atmospheric operation decreases cost at smaller scale, and the BCL has some commercial experience (demo in Burlington, VT (Paisley et al. 1998)). Because biomass gasification temperatures are relatively low, significant depar-tures from equilibrium are found in the product gas. Therefore, kinetic gasifier modelling is complex and different for each reactor type (Consonni et al. 1994;

Li et al. 2001). The main performance characteristics of both gasifiers are given in Table 2.1.

Oxygen Supply

Gasifiers demand oxygen, provided as air, pure oxygen, or combination of the two. The use of pure oxygen reduces the volume flows through the IGT gasifier and through downstream equipment, which reduces investment costs. Also the Autothermal Reformer (see below) is, for the same reason, preferably fired by oxygen. As the production of oxygen is expensive, there will likely be an economical optimum in oxygen purity. Oxygen-enriched air could be a compromise between a cheaper oxygen supply and a reduced downstream equipment size.

Cryogenic air separation is commonly applied when large amounts of O_2 (over 1000 Nm^3/h) are required. Since air is freely available, the costs for oxygen production are directly related to the costs for air compression and refrigeration, the main unit operations in an air separation plant. As a consequence, the oxygen price is mainly determined by the energy costs and plant investment costs (van Dijk et al. 1995; van Ree 1992).

The conventional air separation unit is both capital and energy intensive. A potential for cost reduction is the development of air separation units based on conductive ionic transfer membranes (ITM) that operate on the partial pressure differential of oxygen to passively produce pure oxygen. Research and development of the ITM are in the demonstration phase (DeLallo et al. 2000). Alternative options are membrane air separation, sorption technologies, and water decomposition, but these are less suitable for large-scale application (van Ree 1992).

GAS CLEANING AND CONTAMINANT LIMITS

Raw Gas versus System Requirements

The raw synthesis gas produced by gasification contains impurities. The most typical impurities are organic impurities like condensable tars, BTX (benzene, toluene, and xylenes), inorganic impurities (NH_3, HCN, H_2S, COS, and HCl), volatile metals, dust, and soot (Tijmensen 2000; van Ree et al. 1995). These contaminants can lower catalyst activity in reformer, shift, and methanol reactor, and cause corrosion in compressors, heat exchangers and the (optional) gas turbine.

The estimated maximal acceptable contaminant concentrations are summarized in Table 2.2 together with the effectiveness of wet and dry gas cleaning, as described below.

The gas can be cleaned using available conventional technology, by applying gas cooling, low-temperature filtration, and water scrubbing at 100–250°C. Alternatively, hot gas cleaning can be considered, using ceramic filters and reagents at 350–800°C. These technologies have been described thoroughly by several authors (Consonni et al. 1994; Kurkela 1996; Tijmensen 2000; van Dijk et al. 1995; van Ree et al. 1995). The considered pressure range is no problem for

TABLE 2.1
Characteristics of Gasifiers

	IGT[6] Bubbling Fluidized Bed		IGT max H$_2$[7] Bubbling Fluidized Bed		BCL[8] Indirectly Heated Fast Fluidized Bed	
Biomass input dry basis[1] (tonne/hr)	80		80		80	
Initial moisture content (%)	30		30		30	
Dry moisture content (%)	15		15		10	
HHV$_{dry}$ biomass (GJ/tonne)	19.28		19.28		19.46	
LHV$_{wet}$ biomass[2] (GJ/tonne)	11.94		11.94		12.07	
Steam demand drier[3] (tonne/hr)	26.2		26.2 tonne/hr		33.0 tonne/hr	
Thermal biomass input (MW)	HHV 428.4 / LHV 379.0		HHV 428.4 / LHV 379.0		HHV 432.4 / LHV 383.2	
Steam (kg/kg dry feed)	0.3		0.8		0.019	
Steam[4] (tonne/hr)	24		64		1.52	
Oxygen (kg/kg dry feed)	0.3		0.38		0	
Air (kg/kg dry feed)	0		0		2.06	
Product temperature (°C)	982		920		863	
Exit pressure (bar)	34.5		25		1.2	
Gas yield (kmol/dry tonne)	82.0		121[5]		45.8	
Wet gas output kmol/hour	6560		9680		3664	
Composition: mole fraction on wet basis (on dry basis)						
H$_2$O	0.318	(–)	0.48	(–)	0.199	(–)
H$_2$	0.208	(0.305)	0.24	(0.462)	0.167	(0.208)
CO	0.15	(0.22)	0.115	(0.221)	0.371	(0.463)
CO$_2$	0.239	(0.35)	0.16	(0.308)	0.089	(0.111)
CH$_4$	0.0819	(0.12)	0.005	(0.009)	0.126	(0.157)
C$_2$H$_4$	0.0031	(0.005)	0		0.042	(0.052)
C$_2$H$_6$	0		0		0.006	(0.0074)
O$_2$	0		0		0	
N$_2$	0		0		0	
	1	(1)	1	(1)	1	(1)
LHV$_{wet}$ synthesis gas (MJ/Nm^3)	6.70		3.90		12.7	
Thermal flow (MW)	HHV 352 / LHV 296		HHV 309 / LHV 231		HHV 348 / LHV 316	

TABLE 2.1 (CONTINUED)
Characteristics of Gasifiers

[1] 640 ktonne dry wood annual, load is 8000 h.

[2] Calculated from $LHV_{wet} = HHV_{dry} \times (1 - W) - E_w \times (W + H_{wet} \times m_{H2O})$; with E_w the energy needed for water evaporation (2.26 MJ/kg), H_{wet} the hydrogen content on wet basis (for wood H_{dry} = 0.062) and m_{H2O} the amount of water created from hydrogen (8.94 kg/kg).

[3] Wet biomass: 80/0.7 = 114 tonne/hr to dry biomass 80/0.85 = 94.1 tonne/hr for IGT Π evaporate water 20.2 tonne/hr at 1.3 ts/twe in Niro (indirect) steam dryer. Calculation for BCL is alike. The steam has a pressure of 12 bar and a temperature of minimally 200°C (Pierik et al. 1995).

[4] Pressure is 34.5, 25, or 1.2 bar, temperature is minimally 250, 240, or 120°C.

[5] Calculated from the total mass stream, 188.5 tonne/hr.

[6] Quoted from OPPA (1990) by Williams et al. (1995).

[7] Knight (1998).

[8] Compiled by Williams et al. (1995).

either of the technologies. Hot gas cleaning is advantageous for the overall energy balance when a reformer or a ceramic membrane is applied directly after the cleaning section, because these processes require a high inlet temperature. However, not all elements of hot gas cleaning are yet proven technology, while there is little uncertainty about the cleaning effectiveness of low temperature gas cleaning. Both cleaning concepts are depicted in Figure 2.4.

Tar Removal

Especially in atmospheric gasification, larger hydrocarbons are formed, generally categorized as "tars." When condensing, they foul downstream equipment, coat surfaces, and enter pores in filters and sorbents. To avoid this, their concentration throughout the process must be below the condensation point. On the other hand, they contain a lot of potential CO and H_2. They should thus preferably be cracked into smaller hydrocarbons. Fluidized beds produce tar at about 10 g/m_{NTP}^3 or 1–5 wt% of the biomass feed (Boerrigter et al. 2003; Milne et al. 1998; Tijmensen 2000). BTX, accounting for 0.5 volume % of the synthesis gas, have to be removed prior to the active carbon filters, which otherwise sorb the BTX completely and quickly get filled up (Boerrigter et al. 2003).

Three methods may be considered for tar removal/cracking: thermal cracking, catalytic cracking, and scrubbing. At temperatures above 1000–1200°C, tars are destroyed without a catalyst, usually by the addition of steam and oxygen, which acts as a selective oxidant (Milne et al. 1998). Drawbacks are the need for expensive materials, the soot production, and the low thermal efficiency. Catalytic cracking (dolomite or Ni based) is best applied in a secondary bed and avoids the mentioned problems of thermal cracking. However, the technology is not yet fully proven (Milne et al. 1998). It is not clear to what extent tars are removed (Tijmensen 2000) and the catalyst consumption and costs are matters of concern.

TABLE 2.2

Estimated Contaminant Specifications for Methanol Synthesis[1] and Cleaning Effectiveness of Wet and Dry Gas Cleaning

Contaminant	Gas Phase Specification	Treatment Method and Remarks	
		Existing Technologies	Dry Gas Cleaning[3]
Soot (dust, char, ash)	0 ppb	Cyclones, metal filters, moving beds, candle filters, bag filters, special soot scrubber. Specifications are met.	
Alkaline (halide) metals	<10 ppb	Active coal bed meets specification.[2]	Sorbents under development.
Tar	Below dew point	All tar and BTX: Thermal tar cracker, Oil scrubber,[4] Specifications are met.	All tar and BTX: Catalytic tar cracker, other catalytic operations.
	Catalyst poisoning compounds <1 ppmV		
BTX	Below dew point		Under development.
Halide compounds			
HCl (HBr, HF)	<10 ppb	Removed by aqueous scrubber. Active coal bed meets specification. Absorbed by dolomite in tar cracker (if applicable).	In-bed sorbents or in-stream sorbents. <1 ppm. Guardbeds necessary.
Nitrogen compounds	Total N < 1 ppmV		All nitrogen:
NH_3		Removed by aqueous scrubber. Removed to specification.	catalytic decomposition, combined removal of NH_3/H_2S.
HCN		Active coal bed possibly preceded by hydrolysis to NH_3. Specifications are met.	Selective oxidation under development.
Sulfur compounds	Total S < 1 ppmV[2]		All sulfur:
H_2S		ZnO guard bed in case of high sulfur loads a special removal step, *e.g.*, Claus unit.	In-bed calcium sorbents. Metal oxide sorbents <20 ppm.
COS		Active coal bed possibly preceded by hydrolysis to H_2S. Specifications are met.	

TABLE 2.2 (CONTINUED)
Estimated Contaminant Specifications for Methanol Synthesis[1] and Cleaning Effectiveness of Wet and Dry Gas Cleaning

[1] Most numbers are quoted from Fischer-Tropsch synthesis over a cobalt catalyst (Bechtel 1996; Boerrigter et al. 2003; Tijmensen 2000). Gas turbine specifications are met when FT specifications are.
[2] Cleaning requirements for MeOH synthesis are 0.1 (van Dijk et al. 1995) to 0.25 ppm H_2S (Katofsky 1993). Total sulfur <1 ppmV (Boerrigter et al. 2003). For Fischer-Tropsch synthesis requirements are even more severe: 10 ppb (Tijmensen 2000).
[3] Hot gas cleaning was practiced in the Värnamo Demonstration plant, Sweden (Kwant 2001). All data on dry gas cleaning here is based on the extensive research into high-temperature gas cleaning by Mitchell (Mitchell 1997; Mitchell 1998).
[4] Bergman et al. (Bergman et al. 2003).

Per kg dry wood (15% moisture), 0.0268 kg dolomite. Part of the H_2S and HCl present adsorb on dolomite (van Ree et al. 1995). The tar crackers can be integrated with the gasifier.

Tars can also be removed at low temperature by advanced scrubbing with an oil-based medium (Bergman et al. 2003; Boerrigter et al. 2003). The tar is subsequently stripped from the oil and reburned in the gasifier. At atmospheric pressures BTX are only partially removed, about 6 bar BTX are fully removed. The gas enters the scrubber at about 400°C, which allows high-temperature heat exchange before the scrubber.

Wet Gas Cleaning

When the tars and BTX are removed, the other impurities can be removed by standard wet gas cleaning technologies or advanced dry gas cleaning technologies.

Wet low-temperature synthesis gas cleaning is the preferred method for the short term (van Ree et al. 1995). This method will have some energy penalty and requires additional waste water treatment, but in the short term it is more certain to be effective than hot dry gas cleaning.

A cyclone separator removes most of the solid impurities, down to sizes of approximately 5 μm (Katofsky 1993). New generation bag filters made from glass and synthetic fibers have an upper temperature limit of 260°C (Perry et al. 1987). At this temperature particulates and alkali, which condense on particulates, can successfully be removed (Alderliesten 1990; Consonni et al. 1994; Tijmensen 2000; van Ree et al. 1995). Before entering the bag filter, the synthesis gas is cooled to just above the water dew point.

After the filter unit, the synthesis gas is scrubbed down to 40°C below the water dew point, by means of water. Residual particulates, vapor phase chemical species (unreacted tars, organic gas condensates, trace elements), reduced halogen

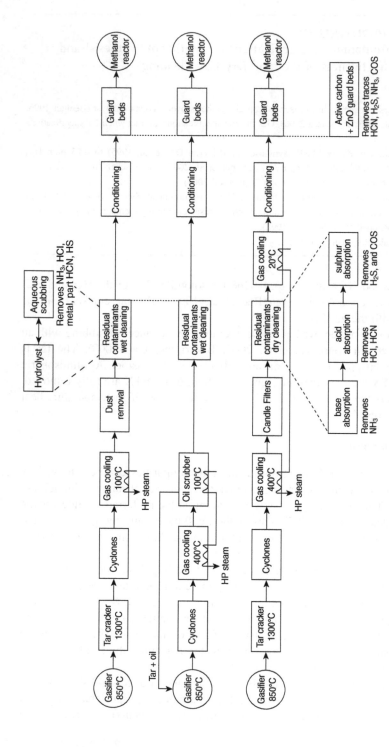

FIGURE 2.4 Three possible gas cleaning trains. Top: tar cracking and conventional wet gas cleaning; middle: tar scrubbing and conventional wet gas cleaning; and bottom: tar cracking and dry gas cleaning.

gases and reduced nitrogen compounds are removed to a large extent. The scrubber can consist of a caustic part where the bulk of H_2S is removed using a NaOH solution (van Ree et al. 1995) and an acid part for ammonia/cyanide removal. Alkali removal in a scrubber is essentially complete (Consonni et al. 1994).

With less than 30 ppm H_2S in the biomass derived synthesis gas, a ZnO bed may be sufficient to lower the sulfur concentration below 0.1 ppm. ZnO beds can be operated between 50 and 400°C, the high-end temperature favors efficient utilization. At low temperatures and pressures, less sulfur is absorbed; therefore, multiple beds will be used in series. The ZnO bed serves one year and is not regenerated (Katofsky 1993; van Dijk et al. 1995). Bulk removal of sulfur is thus not required, but if CO_2 removal is demanded as well (see page 23), a solvent absorption process like Rectisol or Sulfinol could be placed downstream, which also removes sulfur. H_2S and COS are reduced to less than 0.1 ppm and all or part of the CO_2 is separated (*Hydrocarbon Processing* 1998).

Dry/Hot Gas Cleaning

In dry/hot gas cleaning, residual contaminations are removed by chemical absorbents at elevated temperature. In the methanol process, hot gas cleaning has few *energy* advantages as the methanol reactor operates at 200–300°C, especially when preceding additional compression is required (efficient compression requires a cold inlet gas). However, dry/hot gas cleaning may have lower *operational* costs than wet gas cleaning (Mitchell 1998). Within ten years hot gas cleaning may become commercially available for BIG/CC applications (Mitchell 1998). However, requirements for methanol production, especially for catalyst operation, are expected to be more severe (Tijmensen 2000). It is not entirely clear to what extent hot gas cleaning will be suitable in the production of methanol.

Tars and oils are not expected to be removed during the hot gas cleaning since they do not condense at high temperatures. Therefore, they must be removed prior to the rest of the gas cleaning, as discussed above.

For particle removal at temperatures above 400°C, sliding granular bed filters are used instead of cyclones. Final dust cleaning is done using ceramic candle filters (Klein Teeselink et al. 1990; Williams 1998) or sintered-metal barriers operating at temperatures up to 720°C; collection efficiencies greater that 99.8% for 2–7 µm particles have been reported (Katofsky 1993). Still better ceramic filters for simultaneous SO_x, NO_x, and particulate removal are under development (White et al. 1992).

Processes for alkali removal in the 750–900°C range are under development and expected to be commercialized within a few years. Lead and zinc are not removed at this temperature (Alderliesten 1990). High-temperature alkali removal by passing the gas stream through a fixed bed of sorbent or other material that preferentially adsorbs alkali via physical adsorption or chemisorption was discussed by Turn et al. (1998). Below 600°C alkali metals condense onto particulates and can more easily be removed with filters (Katofsky 1993).

Nickel-based catalysts have proved to be very efficient in decomposing tar, ammonia, and methane in biomass gasification gas mixtures at about 900°C. However, sulfur can poison these catalysts (Hepola et al. 1997; Tijmensen 2000). It is unclear if the nitrogenous component HCN is removed. It will probably form NO_x in a gas turbine (Verschoor et al. 1991).

Halogens are removed by sodium and calcium-based powdered absorbents. These are injected in the gas stream and removed in the dedusting stage (Verschoor et al. 1991).

Hot gas desulfurization is done by chemical absorption to zinc titanate or iron oxide-on-silica. The process works optimally at about 600°C or 350°C, respectively. During regeneration of the sorbents, SO_2 is liberated and has to be processed to H_2SO_4 or elemental sulfur (Jansen 1990; Jothimurugesan et al. 1996). ZnO beds operate best close to 400°C (van Dijk et al. 1995).

Early compression would reduce the size of gas cleaning equipment. However, sulfur and chloride compounds condense when compressed and they may corrode the compressor. Therefore, intermediate compression to 6 bar takes place only after bulk removal and 60 bar compression just before the guardbed.

GAS CONDITIONING

Reforming

The synthesis gas can contain a considerable amount of methane and other light hydrocarbons, representing a significant part of the heating value of the gas. Steam reforming (SMR) converts these compounds to CO and H_2 driven by steam addition over a catalyst (usually nickel) at high temperatures (Katofsky 1993). Autothermal reforming (ATR) combines partial oxidation in the first part of the reactor with steam reforming in the second part, thereby optimally integrating the heat flows. It has been suggested that ATR, due to a simpler concept, could become cheaper than SMR (Katofsky 1993), although others suggest much higher prices (Oonk et al. 1997). There is dispute on whether the SMR can deal with the high CO and C+ content of the biomass synthesis gas. While Katofsky writes that no additional steam is needed to prevent coking or carbon deposition in SMR, Tijmensen (2000) poses that this problem does occur in SMR and that ATR is the only technology able to prevent coking.

Steam reforming is the most common method of producing a synthesis gas from natural gas or gasifier gas. The highly endothermic process takes place over a nickel-based catalyst:

$$CH_4 + H_2O \rightarrow CO + 3H_2 \tag{2.1}$$

$$C_2H_4 + 2H_2O \rightarrow 2CO + 4H_2 \tag{2.2}$$

$$C_2H_6 + 2H_2O \rightarrow 2CO + 5H_2 \tag{2.3}$$

Concurrently, the water gas shift reaction (see below) takes place and brings the reformer product to chemical equilibrium (Katofsky 1993).

Reforming is favored at lower pressures, but elevated pressures benefit economically (smaller equipment). Reformers typically operate at 1–3.5 MPa. Typical reformer temperature is between 830°C and 1000°C. High temperatures do not lead to a better product mix for methanol production (Katofsky 1993). The inlet stream is heated by the outlet stream up to near the reformer temperature to match reformer heat demand and supply. In this case less synthesis gas has to be burned compared to a colder gas input, this eventually favors a higher methanol production. Although less steam can be raised from the heat at the reformer outlet, the overall efficiency is higher.

SMR uses steam as the conversion reactant and to prevent carbon formation during operation. Tube damage or even rupture can occur when the steam-to-carbon ratio drops below acceptable limits. The specific type of reforming catalyst used, the operating temperature, and the operating pressure are factors that determine the proper steam-to-carbon ratio for a safe, reliable operation. Typical steam to hydrocarbon-carbon ratios range from 2.1 for natural gas feeds with CO_2 recycle, to 3:1 for natural gas feeds without CO_2 recycle, propane, naphtha, and butane feeds (King et al. 2000). Usually full conversion of higher hydrocarbons in the feedstock takes place in an adiabatic prereformer. This makes it possible to operate the tubular reformer at a steam-to-carbon ratio of 2.5. When higher hydrocarbons are still present, the steam-to-carbon ratio should be higher: 3:5. In older plants, where there is only one steam reformer, the steam-to-carbon ratio was typically 5.5. A higher steam:carbon ratio favors a higher H_2CO ratio and thus higher methanol production. However, more steam must be raised and heated to the reaction temperature, thus decreasing the process efficiency. Neither is additional steam necessary to prevent coking (Katofsky 1993).

Preheating the hydrocarbon feedstock with hot flue gas in the SMR convection section, before steam addition, should be avoided. Dry feed gas must not be heated above its cracking temperature. Otherwise, carbon may be formed, thereby decreasing catalyst activities, increasing pressure drop, and limiting plant throughput. In the absence of steam, cracking of natural gas occurs at temperatures above 450°C, while the flue gas exiting SMRs is typically above 1000°C (King et al. 2000).

Nickel catalysts are affected by sulfur at concentrations as low as 0.25 ppm. An alternative would be to use catalysts that are resistant to sulfur, such as sulphided cobalt/molybdate. However, since other catalysts downstream of the reformer are also sensitive to sulfur, it makes the most sense to remove any sulfur before conditioning the synthesis gas (Katofsky 1993). The lifetime of catalysts ranges from 3 years (van Dijk et al. 1995) to 7 years (King et al. 2000). The reasons for change out are typically catalyst activity loss and increasing pressure drop over the tubes.

Autothermal reforming (ATR) combines steam reforming with partial oxidation. In ATR, only part of the feed is oxidized, enough to supply the necessary heat to steam reform the remaining feedstock. The reformer produces a synthesis gas with a lower H_2.CO ratio than conventional steam methane reforming (Katofsky 1993; Pieterman 2001).

An Autothermal Reformer consists of two sections. In the burner section, some of the preheated feed/steam mixture is burned stoichiometrically with oxygen to produce CO_2 and H_2O. The product and the remaining feed are then fed to the reforming section that contains the nickel-based catalyst (Katofsky 1993).

With ATR, considerably less synthesis gas is produced, but also considerably less steam is required due to the higher temperature. Increasing steam addition hardly influences the H_2:CO ratio in the product, while it does dilute the product with H_2O (Katofsky 1993). Typical ATR temperature is between 900°C and 1000°C.

Since autothermal reforming does not require expensive reformer tubes or a separate furnace, capital costs are typically 50–60% less than conventional steam reforming, especially at larger scales (Dybkjaer et al. 1997, quoted by Pieterman 2001). This excludes the cost of oxygen separation. ATR could therefore be attractive for facilities that already require oxygen for biomass gasification (Katofsky 1993).

The major source of H_2 in oil refineries, catalytic reforming, is decreasing. The largest quantities of H_2 are currently produced from synthesis gas by steam-reforming of methane, but this approach is both energy and capital intensive. Partial oxidation of methane with air as the oxygen source is a potential alternative to the steam-reforming processes. In methanol synthesis starting from C_1 to C_3, it offers special advantages. The amount of methanol produced per kmol hydrocarbon may be 10% to 20% larger than in a conventional process using a steam reformer (de Lathouder 1982). However, the large dilution of product gases by N_2 makes this path uneconomical, and, alternatively, use of pure oxygen requires expensive cryogenic separation (Maiya et al. 2000).

Reforming is still subject to innovation and optimization. Pure oxygen can be introduced in a partial oxidation reactor by means of a ceramic membrane, at 850–900°C, in order to produce a purer synthesis gas. Lower temperature and lower steam to CO ratio in the shift reactor leads to a higher thermodynamic efficiency while maximizing H_2 production (Maiya et al. 2000).

Water Gas Shift

The synthesis gas produced by the BCL and IGT gasifiers has a low H_2:CO ratio. The water gas shift (WGS) reaction (Equation 2.4) is a common process operation to shift the energy value of the carbon monoxide to the hydrogen, which can then be separated using pressure swing adsorption. If the stoichiometric ratio of H_2, CO, and CO_2 is unfavorable for methanol production, the water gas shift can be used in combination with a CO_2 removal step. The equilibrium constant for the WGS increases as temperature decreases. Hence, to increase the production to H_2 from CO, it is desirable to conduct the reaction at lower temperatures, which is also preferred in view of steam economy. However, to achieve the necessary reaction kinetics, higher temperatures are required (Armor 1998; Maiya et al. 2000).

$$CO + H_2O \leftrightarrow CO_2 + H_2 \qquad (2.4)$$

The water gas shift reaction is exothermic and proceeds nearly to completion at low temperatures. Modern catalysts are active at temperatures as low as 200°C (Katofsky 1993) or 400°C (Maiya et al. 2000). Due to high-catalyst selectivity, all gases except those involved in the water–gas shift reaction are inert. The reaction is independent of pressure.

Conventionally, the shift is realized in a successive high temperature (360°C) and low temperature (190°C) reactor. Nowadays, the shift section is often simplified by installing only one CO-shift converter operating at medium temperature (210°C) (Haldor Topsoe 1991). For methanol synthesis, the gas can be shifted partially to a suitable H_2:CO ratio; therefore, "less than one" reactor is applied. The temperature may be higher because the reaction needs not to be complete and this way less process heat is lost.

Theoretically the steam:carbon monoxide ratio could be 2:1. On a lab scale good results are achieved with this ratio (Maiya et al. 2000). In practice extra steam is added to prevent coking (Tijmensen 2000).

CO_2 Removal

The synthesis gas from the gasifier contains a considerable amount of CO_2. After reforming or shifting, this amount increases. To get the ratio $(H_2–CO_2)/(CO + CO_2)$ to the value desired for methanol synthesis, part of the carbon dioxide could be removed. For this purpose, different physical and chemical processes are available. Chemical absorption using amines is the most conventional and commercially best-proven option. Physical absorption, using Selexol, has been developed since the seventies and is an economically more attractive technology for gas streams containing higher concentrations of CO_2. As a result of technological development, the choice for one technology or another could change in time, e.g., membrane technology, or still better amine combinations, could play an important role in future.

Chemical absorption using amines is especially suitable when CO_2 partial pressures are low, around 0.1 bar. It is a technology that makes use of chemical equilibria, shifting with temperature rise or decline. Basically, CO_2 binds chemically to the absorbent at lower temperatures and is later stripped off by hot steam. Commonly used absorbents are alkanolamines applied as solutions in water. Alkanolamines can be divided into three classes: primary, secondary, and tertiary amines. Most literature is focused on primary amines, especially monoethanolamine (MEA), which is considered the most effective in recovering CO_2 (Farla et al. 1995; Wilson et al. 1992), although it might well be that other agents are also suitable as absorbents (Hendriks 1994). The Union Carbide "Flue Guard" process and the Fluor Daniel Econamine FG process (formerly known as the

Dow Chemical Gas/Spec FT-1 process) use MEA, combined with inhibitors to reduce amine degradation and corrosion. The cost of amine-based capture are determined by the cost of the installation, the annual use of amines, the steam required for scrubbing and the electric power. There is influence of scale and a strong dependence on the CO_2 concentration (Hendriks 1994). The investment costs are inversely proportional to the CO_2 concentration in the feed gas when these range from 4% to 8%. MEA is partly entrained in the gas phase; this results in chemical consumption of 0.5–2 kg per tonne CO_2 recovered (Farla et al. 1995; Suda et al. 1992). The presence of SO_2 leads to an increased solvent consumption (Hendriks 1994).

When the CO_2 content makes up an appreciable fraction of the total gas stream, the cost of removing it by heat regenerable reactive solvents may be out of proportion compared to the value of the CO_2. To overcome the economic disadvantages of heat-regenerable processes, physical absorption processes have been developed that are based on the use of essentially anhydrous organic solvents, which dissolve the acid gases and can be stripped by reducing the acid–gas partial pressure without the application of heat. Physical absorption requires a high partial pressure of CO_2 in the feed gas to be purified, 9.5 bar is given as an example by Hendriks (1994). Most physical absorption processes found in the literature are Selexol, which is licensed by Union Carbide, and Lurgi's Rectisol (Hendriks 1994; Hydrocarbon Processing 1998; Riesenfeld et al. 1974). These processes are commercially available and frequently used in the chemical industry. In a countercurrent flow absorption column, the gas comes into contact with the solvent, a 95% solution of the dimethyl ether of polyethylene glycol in water. The CO_2 rich solvent passes a recycle flash drum to recover co-absorbed CO and H_2. The CO_2 is recovered by reducing the pressure through expanders. This recovery is accomplished in serially connected drums. The CO_2 is released partly at atmospheric pressure. After the desorption stages, the Selexol still contains 25–35% of the originally dissolved CO_2. This CO_2 is routed back to the absorber and is recovered in a later cycle. The CO_2 recovery rate from the gas stream will be approximately 98% to 99% when all losses are taken into account. Half of the CO_2 is released at 1 bar and half at elevated pressure: 4 bar. Minor gas impurities such as carbonyl sulfide, carbon disulfide and mercaptans are removed to a large extent, together with the acid gases. Also hydrocarbons above butane are largely removed. Complete acid–gas removal, i.e., to ppm level, is possible with physical absorption only, but is often achieved in combination with a chemical absorption process. Selexol can also remove H_2S, if this were not done in the gas-cleaning step.

It has been suggested by De Lathouder (1982) to scrub CO_2 using crude methanol from the synthesis reactor that has not yet been expanded. The pressure needed for the CO_2 absorption into the methanol is similar to the methanol pressure directly after synthesis. This way only a limited amount of CO_2 is removed, and the required CO_2 partial pressure is high, but the desired R can be reached if conditions are well chosen. The advantage of this method is that no separate regeneration step is required and that it is not necessary to apply extra cooling of the gas stream before the scrubbing operation. The CO_2 loaded crude methanol can be expanded to about atmospheric pressure, so that the carbon dioxide is again released, after which the methanol is purified as would normally be the case.

Physical adsorption systems are based on the ability of porous materials (e.g., zeolites) to selectively adsorb specific molecules at high pressure and low temperature and desorb them at low pressure and high temperature. These processes are already commercially applied in hydrogen production, besides a highly pure hydrogen stream a pure carbon dioxide stream is coproduced. Physical adsorption technologies are not yet suitable for the separation of CO_2 only, due to the high energy consumption (Ishibashi et al. 1998; Katofsky 1993).

METHANOL SYNTHESIS

Methanol is produced by the hydrogenation of carbon oxides over a suitable (copper oxide, zinc oxide, or chromium oxide-based) catalyst:

$$CO + 2H_2 \leftrightarrow CH_3OH \qquad (2.5)$$

$$CO_2 + 3H_2 \leftrightarrow CH_3OH + H_2O \qquad (2.6)$$

The first reaction is the primary methanol synthesis reaction, a small amount of CO_2 in the feed (2–10%) acts as a promoter of this primary reaction and helps maintain catalyst activity. The stoichiometry of both reactions is satisfied when R in the following relation is 2.03 minimally (Katofsky 1993). H_2 builds up in the recycle loop; this leads to an actual R value of the combined synthesis feed (makeup plus recycle feed) of 3 to 4 typically.

$$R = \frac{H_2 - CO_2}{CO + CO_2} \qquad (2.7)$$

The reactions are exothermic and give a net decrease in molar volume. Therefore, the equilibrium is favored by high pressure and low temperature. During production, heat is released and has to be removed to keep optimum catalyst life and reaction rate. 0.3% of the produced methanol reacts further to form side products such as dimethyl ether, formaldehyde, or higher alcohols (van Dijk et al. 1995).

The catalyst deactivates primarily because of loss of active copper due to physical blockage of the active sites by large by-product molecules; poisoning by halogens or sulfur in the synthesis gas, which irreversibly form inactive copper salts; and sintering of the copper crystallites into larger crystals, which then have a lower surface area-to-volume ratio.

Conventionally, methanol is produced in two-phase systems, the reactants and products forming the gas phase and the catalyst forming the solid phase. The production of methanol from synthesis gas was first developed at BASF in Germany in 1922. This process used a zinc oxide/chromium oxide catalyst with poor selectivity, and required extremely vigorous conditions—pressures ranging from 300–1000 bar and temperatures of about 400°C. In the 1960s and 1970s the more active Cu/Zn/Al catalyst was developed allowing more energy-efficient and cost-effective plants, and larger scales. Processes under development at present focus on shifting the equilibrium to the product side to achieve higher conversion per pass. Examples are the gas/solid/solid trickle flow reactor, with a fine adsorbent powder flowing down a catalyst bed and picking up the produced methanol, and liquid phase methanol processes where reactants, product, and catalyst are suspended in a liquid. Fundamentally different could be the direct conversion of methane to methanol, but despite a century of research this method has not yet proved advantageous.

Fixed-Bed Technology

Two reactor types predominate in plants built after 1970 (Cybulski 1994; Kirk-Othmer 1995). The ICI low-pressure process is an adiabatic reactor with cold unreacted gas injected between the catalyst beds (Figure 2.5, left). The subsequent heating and cooling leads to an inherent inefficiency, but the reactor is very reliable and therefore still predominant. The Lurgi system (Figure 2.5, right), with the catalyst loaded into tubes and a cooling medium circulating on the outside of the tubes, allows near-isothermal operation. Conversion to methanol is limited by equilibrium considerations and the high temperature sensitivity of the catalyst. Temperature moderation is achieved by recycling large amounts of hydrogen-rich gas, utilizing the higher heat capacity of H_2 gas and the higher gas velocities to enhance the heat transfer. Typically a gas phase reactor is limited to about 16% CO gas in the inlet to the reactor, in order to limit the conversion per pass to avoid excess heating.

The methanol synthesis temperature is typically between 230 and 270°C. The pressure is between 50 and 150 bar. Higher pressures give an economic benefit,

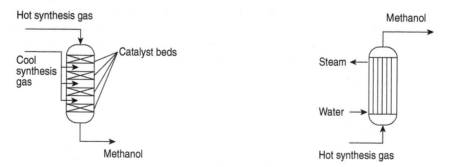

FIGURE 2.5 Methanol reactor types: adiabatic quench (left) and isothermal steam raising (right).

since the equilibrium then favors methanol. Only a part of the CO in the feed gas is converted to methanol in one pass through the reactor, due to the low temperature at which the catalyst operates. The unreacted gas is recycled at a ratio typically between 2.3 and 6.

The copper catalyst is poisoned by both sulfur and chlorine, but the presence of free zinc oxides does help prevent poisoning.

Liquid-Phase Methanol Production

In liquid-phase processes (Cybulski 1994; USDOE 1999), the heat transfer between the solid catalyst and the liquid phase is highly efficient, and therefore the process temperature is very uniform and steady. A gas phase delivers reactants to the finely divided catalyst and removes the products swiftly. This allows high conversions to be obtained without loss of catalyst activity. The higher conversion per pass (compared to fixed-bed technology) eliminates the need for a recycle loop, which implies less auxiliary equipment, fewer energy requirements, smaller volumetric flow through the reactor (Katofsky 1993). An additional advantage is the ability to withdraw a spent catalyst and add a fresh catalyst without interrupting the process.

Different reactor types are possible for liquid-phase methanol production, such as fluidized beds and monolithic reactors. Air Products and Chemicals, Inc. invented a slurry bubble column reactor in the late 1970s, which was further developed and demonstrated in the 1980s and 1990s. From 1997 to 2003, a 300-tonne/day demonstration facility at Eastman Chemical Company in Kingsport, TN produced about 400 million liters methanol from coal via gasification (Heydorn et al. 2003).

In the slurry bubble column reactor (Figure 2.6), reactants from the gas bubbles dissolve in the liquid and diffuse to the catalyst surface, where they react. Products then diffuse through the liquid back to the gas phase. Heat is removed by generating steam in an internal tubular heat exchanger.

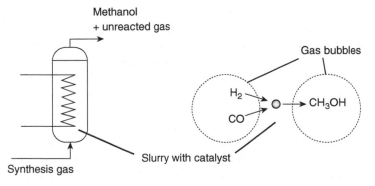

FIGURE 2.6 Liquid phase methanol synthesis with three phases: slurry, gas, and solid.

Commercial Cu/Zn/Al catalysts developed for the two-phase process are used for the three-phase process. The powdered catalyst particles typically measure 1 to 10 μm and are densely suspended in a thermostable oil, chemically resistant to components of the reaction mixture at process conditions, usually paraffin. Catalyst deactivation due to exposure to trace contaminants is a point of concern (Cybulski 1994).

Conversion per pass depends on reaction conditions, catalyst, solvent, and space velocity. Experimental results show 15–40% conversion for CO rich gases and 40–70% CO for balanced and H_2 rich gases. Computation models predict future CO conversions of over 90%, up to 97% respectively (Cybulski 1994; Hagihara et al. 1995). Researchers at the Brookhaven National Laboratory have developed a low temperature (active as low as 100°C) liquid phase catalyst that can convert 90% of the CO in one pass (Katofsky 1993). With steam addition the reaction mixture becomes balanced through the water gas shift reaction. USDOE claims that the initial hydrogen to carbon monoxide ratio is allowed to vary from 0.4 to 5.6 without a negative effect on performance (USDOE 1999).

The investment costs for the liquid-phase methanol process are expected to be 5–23% less than for a gas-phase process of the same methanol capacity. Operating costs are 2–3% lower; this is mainly due to a four times lower electricity consumption (USDOE 1999).

OPTIONS FOR SYNERGY

ELECTRICITY COGENERATION BY COMBINED CYCLE

Unconverted synthesis gas that remains after the methanol production section can still contain a significant amount of chemical energy. These gas streams may be combusted in a gas turbine, although they generally have a much lower heating value (4–10 MJ/m^3_{NTP}) than natural gas or distillate fuel (35–40 MJ/m^3_{NTP}) for which most gas turbine combustors have been designed. When considering commercially available gas turbines for low calorific gas firing, the following items deserve special attention (Consonni et al. 1994; Rodrigues de Souza et al. 2000; van Ree et al. 1995): the combustion stability, the pressure loss through the fuel injection system, and the limits to the increasing mass flow through the turbine.

Different industrial and aeroderivative gas turbines have been operated successfully with low LHV gas, but on the condition that the hydrogen concentration in the gas is high enough to stabilize the flame. Up to 20% H_2 is required at 2.9 MJ/m^3_{NTP}. Hydrogen has a high flame-propagation speed and thus decreases the risk of extinguishing the flame (Consonni et al. 1994).

Injecting a larger fuel volume into the combustor through a nozzle originally designed for a fuel with much higher energy density can lead to pressure losses, and thus to a decreased overall cycle efficiency. Minor modifications are sufficient for most existing turbines. In the longer term, new turbines optimised for low heating value gas might include a complete nozzle combustor redesign (Consonni et al. 1994).

The larger fuel flow rate also implies an increase in mass flow through the turbine expander, relative to natural gas firing. This can be accommodated partly by increasing the turbine inlet pressure, but this is limited by the compressor power available. At a certain moment, the compressor cannot match this increased pressure any more and goes into stall: the compressor blocks. To prevent stall, decreasing the combustion temperature is necessary; this is called derating. This will lower the efficiency of the turbine, though (Consonni et al. 1994; van Ree et al. 1995). Higher turbine capacity would normally give a higher efficiency, but as the derating penalty is also stronger, the efficiency gain is small (Rodrigues de Souza et al. 2000).

Due to the setup of the engine the compressor delivers a specific amount of air. However, to burn one m^3_{NTP} of fuel gas less compressed air is needed compared to firing natural gas. The surplus air can be bled from the compressor at different pressures and used elsewhere in the plant, e.g., for oxygen production (van Ree et al. 1995). If not, efficiency losses occur.

All the possible problems mentioned for the currently available gas turbines can be overcome when designing future gas turbines. Ongoing developments in gas turbine technology increase efficiency and lower the costs per installed kW over time (van Ree et al. 1995). Cooled interstages at the compressor will lower compressor work and produce heat, which can be used elsewhere in the system. Also gas turbine and steam turbine could be put on one axis, which saves out one generator and gives a somewhat higher efficiency.

Turbines set limits to the gas quality. The gas cleaning system needs to match particles and alkali requirements of the gas turbines. When these standards are exceeded, wearing becomes more severe and lifetime and efficiency will drop (van Ree et al. 1995). However, the synthesis gas that passed various catalysts prior to the gas turbine has to meet stricter demands. It is therefore expected that contaminants are not a real problem in gas turbines running on flue gas from methanol production.

NATURAL GAS COFIRING/COFEEDING

If the caloric value of the unconverted synthesis gas is too low for (direct) combustion in a gas turbine, this could be compensated for by cofiring natural gas. Besides raising the heating value of the gas, the application of natural gas can also increase the scale, thermal efficiency, and economics of the gas turbines.

Natural gas can also be applied as cofeeding in the entire process. Or, vice versa, the large scale of existing methanol production units could be utilized by plugging in a biomass gasifier and gas make-up section. The product can be considered partially of biomass origin.

BLACK LIQUOR GASIFICATION

Pulp and paper mills produce huge amounts of black liquor as a residue. They are the most important source of biomass energy in countries such as Sweden and Finland, representing a potential energy source of 250–500 MW per mill. As modern kraft pulp mills have a surplus of energy, they could become key suppliers of renewable fuels in the future energy system, if the primary energy in the black liquor could be converted to an energy carrier of high value.

Ekbom et al. (2003) have evaluated the production of methanol and DME (see below) from black liquor gasification (BLGMF process). This scheme could be realized against reasonable costs, if heat recovery boilers, which economic life has ended, are replaced by BLGMF. Using black liquor as a raw material for methanol/DME production would have the following advantages:

1. Biomass logistics are extremely simplified as the raw material for fuel making is handled within the ordinary operations of the pulp and paper plant.
2. The process is easily pressurized, which enhances fuel production efficiency.
3. The produced syngas has a low methane content, which optimizes fuel yield.
4. Pulp mill economics becomes less sensitive to pulp prices as the economics are diversified with another product.
5. Gasification capital cost is shared between recovery of inorganic chemicals, steam production, and synthesis gas production.

OTHER BIOFUELS VIA GASIFICATION

Gasification, gas cleaning, and make-up are important parts of the process to make methanol from biomass. These parts are also key to the production of hydrogen and Fischer-Tropsch liquids from biomass. Development of methanol from biomass thus offers synergy with development of hydrogen and Fischer-Tropsch liquids. Methanol can also be an intermediate in the production of other renewable fuels such as synthetic diesel, gasoline, and dimethyl ether.

Hydrogen

The production of hydrogen from synthesis gas is somewhat simpler and cheaper than the production of methanol. The gasification step should aim at maximizing the hydrogen yield, which can be further increased by reforming any methane left and a water-gas-shift reaction. Hydrogen separation takes place by pressure swing adsorption or (in future) membranes.

Hydrogen is already produced at large scale in the chemical and oil industry. It is often seen as the future fuel for the transportation sector and households.

Fischer-Tropsch (FT) Diesel

A broad range of hydrocarbons, ranging from methane to waxes of high molecular weight can be produced from synthesis gas using an iron or cobalt catalyst. These are called Fischer-Tropsch (FT) liquids or Gas-to-Liquids (GTL). By cracking the longer hydrocarbons and refining, a diesel is produced that can be blended into standard diesel. FT diesel has very low levels of sulfur and aromatic compounds compared to ordinary diesel and, when processed in an internal combustion engine, emit less NO_x and particulates than diesel fuels.

The FT process was first developed at commercial scale for the production of synthetic oil in Germany during the Second World War and was further developed by Sasol in South Africa. Sasol remains today the only producer of FT products from low grade coals. The rising oil price, availability of large amounts of "stranded gas," and decreasing investment costs have increased the interest in FT liquids. Qatar seems to be the driver of FT development, with planned projects totaling to 800,000 bbl/day or about 114 ktonne/day (Bensaïd 2004).

In 1999 when the world had a considerable surplus of methanol production capacity, several companies proposed to retrofit methanol plants to produce alternative products, e.g., Fischer-Tropsch liquids or hydrogen (Brown 1999; Yakobson 1999). The demonstration unit of Choren in Freiberg Germany, has produced both FT liquids and methanol.

Methanol to Diesel

Lurgi claims to develop a cheaper way to make ultra-clean diesel fuel from synthesis gas via methanol. The process first converts methanol into propylenethesis; this is followed by olefin oligomerization (conversion to distillates), then product separation-plus-hydrogenation. The intermediate methanol-to-propylene step so far is only proven at demonstration scale.

The process would yield mostly kerosene and diesel, along with a small yield of gasoline and light ends. The near-zero sulphur/polyaromatics diesel fuel resulting from this process would differ from more conventional Fischer-Tropsch diesel only in cetane numbe (>52 via "Methanol-to-Synfuel" versus >70 cetane for FT diesel). The incidental gasoline stream not only would be near-zero sulfur, but also have commercial octane ratings (92 RON, 80 MON) and maximally 11% aromatics (Peckham 2003).

Methanol to Gasoline

In the 1970s, Mobil developed and commercialized a methanol to gasoline (MTG) process. A plant was built in Montunui, New Zealand in 1985 and sold to Methanex. It produced gasoline until 1997 when the plant was permanently idled. If the gasoline is to be sold without additional blending, then further treating is necessary to reduce the amount of benzenes.

Dimethyl Ether (DME)

Dimethyl ether (CH_3OCH_3) is generally produced by dehydration of methanol. At large scale, the methanol production and dehydration processes are combined in one reactor, such that the dimethyl ether is produced directly from synthesis gas slightly more efficiently than methanol. The previously mentioned slurry bubble column reactor of Eastman Chemical Company in Kingsport, TN, has been demonstrated to be able to produce DME as well. The LPDME™ Process uses a physical mixture of a commercial methanol catalyst and a commercial dehydration catalyst in a single slurry reactor (Heydorn et al. 2003).

Like methanol, DME has promising features as a fuel candidate for both auto and diesel engines. With small adaptations to engine and fuel system, DME can be used in blends with diesel (10–20%), leading to higher fuel efficiency and lower emissions. In auto engines, DME can be used with LPG (any %) and neat. Since DME is as easily reformed as methanol, it has a big potential as fuel of fuel cell vehicles (van Walwijk et al. 1996). DME can be easily pressurized and handled as a liquid (Ekbom et al. 2003).

TECHNO-ECONOMIC PERFORMANCE

Following the train of components of Figure 2.1 and given the potential options for gasification, gas cleaning and conditioning, synthesis and separation, many routes to produce methanol from biomass can be imagined. The authors have previously analyzed the techno-economic performance of methanol from wood through 6 concepts, which will be recapitulated here. At the end of the section, results will be placed into broader perspective with other literature and with fossil gasoline and diesel.

SELECTION OF CONCEPTS

Some concepts chosen resemble conventional production of methanol from natural gas, making use of wet gas cleaning, steam reforming, shift, and a solid-bed methanol reactor. Similar concepts have previously been analyzed by Katofsky (1993). Advanced components could offer direct or indirect energy benefits (liquid phase-methanol synthesis, hot gas cleaning), or economic benefits (autothermal reforming). Available process units are logically combined so the supplied gas composition of a unit matches the demands of the subsequent unit, and heat leaps are restricted if possible. The following considerations play a role in selecting concepts:

1. The IGT direct oxygen fired pressurized gasifier, in the normal and maximized H_2 option, and the Battelle indirect atmospheric gasifier are considered for synthesis gas production because they deliver a medium calorific nitrogen undiluted gas stream and cover a broad range of gas compositions.

2. Hot gas cleaning is only sensible if followed by *hot* process units like reforming or (intermediate temperature) shifting. Hot gas cleaning is not applied after atmospheric gasification since the subsequent pressurization of the synthesis gas necessitates cooling anyway.

3. For reforming fuel gas produced via an IGT gasifier, an autothermal reformer is chosen, because of higher efficiency and lower costs. The high hydrogen yield possible with steam reforming is less important here since the H_2:CO ratio of the gas is already high. The BCL gasifier, however, is followed by steam reforming to yield more hydrogen.

4. Preceding liquid-phase methanol synthesis, shifting the synthesis gas composition is not necessary since the reaction is flexible toward the gas composition. When steam is added, a shift reaction takes place in the reactor itself. Before gas-phase methanol production the composition is partially shifted and because the reactor is sensitive to CO_2 excess, part of the CO_2 is removed.

5. After the methanol passes through once, the gas still contains a large part of the energy and is expected to suit gas turbine specifications. The same holds for unreformed BCL and IGT gases, which contain energy in the form of C_2+ fractions. When the heating value of the gas stream does not allow stable combustion in a gas turbine, it is fired in a boiler to raise process steam. The chemical energy of IGT+ gas is entirely in hydrogen and carbon monoxide. After once through methanol production, the gas still contains enough chemical energy for combustion in a gas turbine.

6. Heat supply and demand within plants are to be matched to optimize the overall plant efficiency.

7. Oxygen is used as oxidant for the IGT gasifier and the autothermal reformer. The use of air would enlarge downstream equipment size by a factor 4. Alternatively, oxygen-enriched air could be used. This would probably give an optimum between small equipment and low air separation investment costs.

These considerations led to a selection of 6 conversion concepts (see Table 2.3). The six concepts selected potentially have low-cost and/or high-energy efficiency. The concepts are composed making use of both existing commercially available technologies, as well as (promising) new technologies.

MODELING MASS AND ENERGY BALANCES

The selected systems were modeled in Aspen Plus, a widely used process simulation program. In this flowsheeting program, chemical reactors, pumps, turbines, heat exchanging apparatuses, etc. are virtually connected by pipes. Every component can be specified in detail: reactions taking place, efficiencies, dimensions of heating surfaces, and so on. For given inputs, product streams can be calculated, or one can evaluate the influence of apparatus adjustments on electrical output. The plant efficiency can be optimized by matching the heat supply and demand. The resulting dimensions of streams and units and the energy balances can subsequently be used for economic analyses.

TABLE 2.3
Selected Methanol Production Concepts

	Gasifier	Gas Cleaning	Reforming	Shift	Methanol Reactor	Power Generation
1	IGTmaxH$_2$	Wet	–	–	Liquid phase	Combined cycle
2	IGT	Hot (550°C)	ATR	–	Liquid phase, with steam addition	Combined cycle
3	IGT	Wet	–	–	Liquid phase, with steam addition	Combined cycle
4	BCL	Wet	SMR	–	Liquid phase, with steam addition and recycle	Steam cycle
5	IGT	Hot (550°C)	ATR	Partial	Solid bed, with recycle	Steam cycle
6	BCL	Wet	SMR	Partial	Solid bed, with recycle	Steam cycle

The pretreatment and gasification sections are not modelled, their energy use and conversion efficiencies are included in the energy balances, though. The models start with the synthesis gas composition from the gasifiers as given in Table 2.1

The heat supply and demand within the plant is carefully matched and aimed at maximizing the production of superheated steam for the steam turbine. The intention was to keep the integration simple by placing few heat exchangers per gas/water/steam stream. Of course, concepts with more process units demanding more temperature altering are more complex than concepts consisting of few units. First, an inventory of heat supply and demand was made. Streams matching in temperature range and heat demand/supply were combined: e.g., heating before the reformer by using the cooling after the reformer. When the heat demand is met, steam can be raised for power generation. Depending on the amount and ratio of high and low heat, process steam is raised in heat exchangers or drawn from the steam turbine: if there is enough energy in the plant to raise steam of 300°C, but barely superheating capacity, then process steam of 300°C is raised directly in the plant. If there is more superheating than steam-raising capacity, then process steam is drawn from the steam cycle. Steam for gasification and drying is almost always drawn from the steam cycle, unless a perfect match is possible with a heat-supplying stream. The steam entering the steam turbine is set at 86 bar and 510°C.

Table 2.4 summarizes the outcomes of the flowsheet models. In some concepts still significant variations can be made. In concept 4, the reformer needs gas for firing. The reformer can either be entirely fired by purge gas (thus restricting the recycle volume) or by part of the gasifier gas. The first option gives a somewhat higher methanol production and overall plant efficiency. In concept 5, one can choose between a larger recycle and more steam production in the boiler. A recycle of five times the feed volume, instead of four, gives a much higher

TABLE 2.4
Results of the Aspen Plus Performance Calculations for 430-MWth Input HHV Systems (equivalent to 380 MW$_{th}$ LHV for biomass with 30% moisture) of the Methanol Production Concepts Considered

		HHV Output (MW)		
		Fuel	Net Electricity[1]	Efficiency[2]
1	IGT – Max H$_2$, Scrubber, Liquid-Phase Methanol Reactor, Combined Cycle	161	53	50%
2	IGT, Hot Gas Cleaning, Autothermal Reformer, Liquid-Phase Methanol Reactor with Steam Addition, Combined Cycle	173	62	55%
3	IGT, Scrubber, Liquid Phase Methanol Reactor with Steam Addition, Combined Cycle	113	105	51%
4	BCL, Scrubber, Steam Reformer, Liquid-Phase Methanol Reactor with Steam Addition and Recycle, Steam Cycle	246	0	57%
5	IGT, Hot Gas Cleaning, Autothermal Reformer, Partial Shift, Conventional Methanol Reactor with Recycle, Steam Turbine	221	15	55%
6	BCL, Scrubber, Steam Reforming, Partial Shift, Conventional Methanol Reactor with Recycle, Steam Turbine	255	−17	55%

[1] Net electrical output is gross output minus internal use. Gross electricity is produced by gas turbine and/or steam turbine. The internal electricity use stems from pumps, compressors, oxygen separator, etc.

[2] The overall energy efficiency is expressed as the *net overall fuel + electricity efficiency* on an HHV basis. This definition gives a distorted view, since the quality of energy in fuel and electricity is considered equal, while in reality it is not. Alternatively, one could calculate a *fuel only efficiency*, assuming that the electricity part could be produced from biomass at, e.g., 45% HHV in an advanced BIG/CC (Faaij et al. 1998), this definition would compensate for the inequality of electricity and fuel in the most justified way, but the referenced electric efficiency is of decisive importance. Another qualification for the performance of the system could use exergy: the amount of work that could be delivered by the material streams.

methanol production and plant efficiency. Per concept, only the most efficient variation is reported in Table 2.4.

Based on experiences with low calorific combustion elsewhere (Consonni et al. 1994; van Ree et al. 1995), the gas flows in the configurations presented here are expected to give stable combustion in a gas turbine. Table 2.4 only includes the advanced turbines. Advanced turbine configurations, with set high compressor and turbine efficiencies and no dimension restrictions, give gas turbine efficiencies of 41–52% and 1–2% point higher overall plant efficiency than conventional configurations. Based on the overall plant efficiency, the methanol concepts lie in a close range of 50–57%. Liquid-phase methanol production preceded by reforming (concepts 2 and 4) results in somewhat higher overall efficiencies. After the pressurized IGT gasifier hot gas cleaning leads to higher efficiencies than wet gas cleaning, although not better than concepts with wet gas cleaning after a BCL gasifier.

Several units may be realized with higher efficiencies than considered here. For example, new catalysts and carrier liquids could improve liquid-phase methanol single-pass efficiency up to 95% (Hagihara et al. 1995). The electrical efficiency of gas turbines will increase by 2-3% points when going to larger scale (*Gas Turbine World* 1997).

Costing Method

An economic evaluation has been carried out for the concepts considered. Plant sizes of 80, 400, 1000, and 2000 MW_{th} HHV are evaluated, 400 MW_{th} being the base scale. The scale of the conversion system is expected to be an important factor in the overall economic performance. This issue has been studied for BIG/CC systems (Faaij et al. 1998; Larson et al. 1997), showing that the economies of scale of such units can offset the increased costs of biomass transport up to capacities of several hundreds of MW_{th}. The same reasoning holds for the methanol production concepts described here. It should, however, be realized that production facilities of 1000–2000 MW_{th} require very large volumes of feedstock: 200–400 dry tonne/hour or 1.6–3.2 dry Mtonne per year. Biomass availability will be a limitation for most locations for such large-scale production facilities, especially in the shorter term. In the longer term (2010–2030), if biomass production systems become more commonplace, this can change. Very large scale biomass conversion is not without precedent: various large-scale sugar/ethanol plants in Brazil have a biomass throughput of 1–3 Mtonne of sugarcane per year, while the production season covers less than half a year. Also, large paper and pulp complexes have comparable capacities. The base scale chosen is comparable to the size order studied by Williams et al. (1995) and Katofsky (1993), 370–385 MW_{th}.

The methanol production costs are calculated by dividing the total annual costs of a system by the produced amount of methanol. The total annual costs consist of:

1. Annual investments.
2. Operating and maintenance.
3. Biomass feedstock.
4. Electricity supply/demand (fixed power price).

The total annual investment is calculated by a factored estimation (Peters et al. 1980), based on knowledge of major items of equipment as found in the literature or given by experts. The uncertainty range of such estimates is up to ±30%. The installed investment costs for the separate units are added up. The unit investments depend on the size of the components (which follow from the

Aspen Plus modelling), by scaling from known scales in literature (see Table 2.5), using Equation 2.8:

$$\frac{Cost_b}{Cost_a} = \left(\frac{Size_b}{Size_a}\right)^R \qquad (2.8)$$

with R = scaling factor.

Various system components have a maximum size, above which multiple units will be placed in parallel. Hence the influence of economies of scale on the total system costs decreases. This aspect is dealt with by assuming that the base investment costs of multiple units are proportional to the cost of the maximum size: the base investment cost per size becomes constant. The maximum size of the IGT gasifier is subject to discussion, as the pressurised gasifier would logically have a larger potential throughput than the atmospheric BCL.

The total investment costs include auxiliary equipment and installation labour, engineering and contingencies. If only equipment costs, excluding installation, are available, those costs are increased by applying an overall installation factor of 1.86. This value is based on 33% added investment to hardware costs (instrumentation and control 5%, buildings 1.5%, grid connections 5%, site preparation 0.5%, civil works 10%, electronics 7%, and piping 4%) and 40% added installation costs to investment (engineering 5%, building interest 10%, project contingency 10%, fees/overheads/profits 10%, and start-up costs 5%) (Faaij et al. 1998).

The annual investment takes into account the technical and economic lifetime of the installation. The interest rate is 10%.

Operational costs (maintenance, labour, consumables, residual streams disposal) are taken as a single overall percentage (4%) of the total installed investment (Faaij et al. 1998; Larson et al. 1998). Differences between conversion concepts are not anticipated.

It was assumed that enough biomass will be available at 2 US$/GJ (HHV). This is a reasonable price for Latin and North American conditions. Costs of cultivated energy crops in the Netherlands amount approximately 4 US$/GJ and thinnings 3 US$/GJ (Faaij 1997), and biomass imported from Sweden on a large scale is expected to cost 7 US$/GJ (1998). On the other hand biomass grown on Brazilian plantations could be delivered to local conversion facilities at 1.6–1.7 US$/GJ (Hall et al. 1992; Williams et al. 1995). It has been shown elsewhere that international transport of biomass and bioenergy is feasible against modest costs.

Electricity supplied to or demanded from the grid costs 0.03 US$/kWh. The annual load is 8000 hours.

RESULTS

Results of the economic analysis are given in Figure 2.7. The 400 MWth conversion facilities deliver methanol at 8.6–12 US$/GJ. Considering the 30%

TABLE 2.5
Costs of System Components in MUS$_{2001}$ [1]

Unit	Base Investment Cost (fob)	Scale Factor	Base Scale	Overall Installation Factor [22]	Maximum Size [23]
Pretreatment [2]					
Conveyers [3]	0.35	0.8	33.5 wet tonne/hour	1.86 (v)	110
Grinding [3]	0.41	0.6	33.5 wet tonne/hour	1.86 (v)	110
Storage [3]	1.0	0.65	33.5 wet tonne/hour	1.86 (v)	110
Dryer [3]	7.6	0.8	33.5 wet tonne/hour	1.86 (v)	110
Iron removal [3]	0.37	0.7	33.5 wet tonne/hour	1.86 (v)	110
Feeding system [3,4]	0.41	1	33.5 wet tonne/hour	1.86 (v)	110
Gasification System					
BCL [5]	16.3	0.65	68.8 dry tonne/hour	1.69	83
IGT [6]	38.1	0.7	68.8 dry tonne/hour	1.69	75
Oxygen plant (installed) [7]	44.2	0.85	41.7 tonne O_2/hour	1	–
Gas cleaning					
Tar cracker [3]	3.1	0.7	34.2 m^3 gas/s	1.86 (v)	52
Cyclones [3]	2.6	0.7	34.2 m^3 gas/s	1.86 (v)	180
High-temperature heat exchanger [8]	6.99	0.6	39.2 kg steam/s	1.84 (v)	–
Baghouse filter [3]	1.6	0.65	12.1 m^3 gas/s	1.86 (v)	64
Condensing scrubber [3]	2.6	0.7	12.1 m^3 gas/s	1.86 (v)	64
Hot gas cleaning [9]	30	1.0	74.1 m^3 gas/s	1.72 (v)	–
Synthesis Gas Processing					
Compressor [10]	11.1	0.85	13.2 MW$_e$	1.72 (v)	–
Steam reformer [11]	9.4	0.6	1390 kmol total/hour	2.3 (v)	–
Autothermal reformer [12]	4.7	0.6	1390 kmol total/hour	2.3 (v)	–
Shift reactor (installed) [13]	36.9	0.85	15.6 Mmol $CO+H_2$/hour	1	–
Selexol CO_2 removal (installed) [14]	54.1	0.7	9909 kmol CO_2/hour	1	–
Methanol Production					
Gas-phase methanol [15]	7	0.6	87.5 tonne MeOH/hour	2.1 (v)	–
Liquid-phase methanol [16]	3.5	0.72	87.5 tonne MeOH/hour	2.1 (v)	–
Refining [17]	15.1	0.7	87.5 tonne MeOH/hour	2.1 (v)	
Power isle [18]					
Gas turbine + HRSG [3,19]	18.9	0.7	26.3 MW$_e$	1.86 (v)	–

TABLE 2.5 (CONTINUED)
Costs of System Components in MUS$_{2001}$[1]

Unit	Base Investment Cost (fob)	Scale Factor	Base Scale	Overall Installation Factor[22]	Maximum Size[23]
Steam turbine + steam system[3,20]	5.1	0.7	10.3 MW$_e$	1.86 (v)	–
Expansion turbine[21]	4.3	0.7	10.3 MW$_e$	1.86 (v)	–

[1] Annual GDP deflation up to 1994 is determined from OECD (1996) numbers. Average annual GDP deflation after 1994 is assumed to be 2.5% for the United States, 3.0% for the EU. Cost numbers of Dutch origin are assumed to be dependent on the EU market, therefore EU GDP deflators are used. 1_{2001} = 0.94 US$_{2001}$ = 2.204 Dfl$_{2001}$.

[2] Total pretreatment approximately sums up to a base cost of 8.15 MUS$_{2001}$ at a base scale of 33.5 tonne wet/hour with an R factor of 0.79.

[3] Based on first-generation BIG/CC installations. Faaij et al. (1995) evaluated a 29-MW$_e$ BIG/CC installation (input 9.30 kg dry wood/s, produces 10.55 Nm^3 fuel gas/s) using vendor quotes. When a range is given, the higher values are used (Faaij et al. 1998). The scale factors stem from Faaij et al. (1998).

[4] Two double-screw feeders with rotary valves (Faaij et al. 1995).

[5] 12.72 MUS$_{1991}$ (already includes added investment to hardware) for a 1650 dry tonne per day input BCL gasifier, feeding not included, R is 0.7 (Williams et al. 1995). Stronger effects of scale for atmospheric gasifiers (0.6) were suggested by Faaij et al. (1998). Technical director Mr. Paisley of Battelle Columbus, quoted by Tijmensen (2000), estimates the maximum capacity of a single BCL gasifier train at 2000 dry tonnes/day.

[6] 29.74 MUS$_{1991}$ (includes already added investment to hardware) for a 1650 dry tonne/day input IGT gasifier, R = 0.7 (Williams et al. 1995). Maximum input is 400-MW$_{th}$ HHV (Tijmensen 2000).

[7] Air Separation Unit: Plant investment costs are given by Van Dijk (van Dijk et al. 1995): I = $0.1069 \cdot C^{0.8508}$ in MUS$_{1995}$ installed, C = Capacity in tonne O$_2$/day. The relation is valid for 100 to 2000 tonne O$_2$/day. Williams et al. (1995) assume higher costs for small installations, but with a stronger effect of scale: I = $0.260 \cdot C^{0.712}$ in MUS$_{1991}$ fob plus an overall installation factor of 1.75 (25% and 40%). Larson et al. (1998) assume lower costs than Van Dijk, but with an even stronger scaling factor than Williams: 27 MUS$_{1997}$ installed for an 1100 tonne O$_2$ per day plant and R=0.6. We have applied the first formula (by Van Dijk) here. The production of 99.5% pure O$_2$ using an air separation unit requires 250–350 kWh per tonne O$_2$ (van Dijk et al. 1995; van Ree 1992).

[8] High-temperature heat exchangers following the gasifier and (in some concepts) at other locations are modelled as HRSGs, raising steam of 90 bar/520°C. A 39.2-kg steam/s unit costs 6.33 MUS$_{1997}$ fob, overall installation factor is 1.84 (Larson et al. 1998).

[9] Tijmensen (2000) assumes the fob price for hot gas cleaning equipment to be 30 MUS$_{2000}$ for a 400-MW$_{th}$ HHV input. This equals 74.1 m^3/s from a BCL gasifier (T = 863°C, 1.2 bar). There is no effect of scaling.

[10] Katofsky (1993) assumes compressors to cost 700 US$_{1993}$ per required kW$_{mech}$, with an installation factor of 2.1. The relation used here stems from the compressor manufacturer Sulzer quoted by (2000). At the indicated base scale, total installed costs are about 15% higher than assumed by Katofsky. Multiple compressors, for synthesis gas, recycle streams, or hydrogen, are considered as separate units. The overall installation factor is taken 1.72 because the base unit matches a 400-MW$_{th}$ plant rather than a 70-MW$_{th}$ plant.

TABLE 2.5 (CONTINUED)
Costs of System Components in MUS$$_{2001}$$[1]

Unit	Base Investment Cost (fob)	Scale Factor	Base Scale	Overall Installation Factor[22]	Maximum Size[23]

[11] Investments for steam reformers vary from 16.9 MUS$$_{1993}$$, for a throughput of 5800 kmol methane/hour with an overall installation factor of 2.1 (Katofsky 1993) to 7867 k$$_{1995}$$ for a 6.2 kg methane/s (1390 kmol/hour), overall installation factor is 2.3 (van Dijk et al. 1995). These values suggest a strong effect of scaling R = 0.51, while Katofsky uses a modest R = 0.7. Here, we use the values of Van Dijk in combination with an R factor of 0.6. The total amount of moles determines the volume and thus the price of the reactor.

[12] Autothermal reforming could be 50% cheaper than steam reforming (Katofsky 1993), although higher costs are found as well (Oonk et al. 1997).

[13] Investment for shift reactors vary from 9.02 MU$$_{1995}$$ for an 8819 kmol $CO+H_2$/hr reactor, and an overall installation factor is 1.81 (Williams et al. 1995) to 30 MUS$$_{1994}$$ installed for a 350000 Nm³/hr $CO+H_2$/hr (15625 kmol/hr) reactor (Hendriks 1994). Williams assumes an R = 0.65, but comparison of the values suggest only a weak influence of scale (R = 0.94). Here, we use the the values from Hendriks, with R set at 0.85. A dual shift is costed as a shift of twice the capacity.

[14] Costs for CO_2 removal through Selexol amounts 14.3 MUS$$_{1993}$$ fob (overall installation factor is 1.87) for an 810 kmol CO_2/hr unit, R = 0.7 (Katofsky 1993) up to 44 MUS$$_{1994}$$ installed for a 9909 kmol CO_2/hour unit (Hendriks 1994). The value from Hendriks is assumed to be right, since his research into CO_2 removal is comprehensive.

[15] Van Dijk et al. (1995) estimate that a methanol reactor for a 2.1 ktonne methanol per day plant costs 4433 kUS$$_{1995}$$ (fob) or 9526 kUS$$_{1995}$$ installed (overall installation factor is 2.1). The total plant investment in their study is 138 MUS$$_{1995}$$, or 150 MUS$$_{2001}$$. Katofsky (1993) estimates the costs for a 1056 tonne methanol/day plant to be 50 MUS$$_{1995}$$ fob, this excludes the generation and altering of synthesis gas, but includes make-up and recycle compression and refining tower. A 1000 tpd plant costs about 160 MUS$$_{2001}$$, and a 2000 tpd plant 200 MUS$$_{2001}$$, which suggests a total plant scale factor of 0.3 (Hamelinck et al. 2001). These values come near the ones mentioned by Katofsky. This implies that the values given by Van Dijk are too optimistic and should be altered by a factor 1.33. It is therefore assumed that the base investment for the methanol reactor only is 7 MUS$$_{2001}$$, the installation factor is 2.1. The influence of scale on reactor price is not assumed to be as strong as for the complete plant: 0.6.

[16] Installed costs for a 456 tonne per day liquid-phase methanol unit, are 29 MU$$_{1997}$$, excluding generation and altering of synthesis gas, but including make-up and recycle compression, and refining tower. R = 0.72 (Tijm et al. 1997). Corrected for scale and inflation this value is about half the cost of the conventional unit by Katofsky and the corrected costs of Van Dijk. It is therefore assumed that the price of a liquid-phase methanol reactor is 3.5 MUS$$_{2001}$$ for a 2.1 ktonne per day plant, installation factor is 2.1.

[17] Cost number for methanol separation and refining is taken from Van Dijk, increased with 33% as described in note 15.

[18] For indication: A complete combined cycle amounts to about 830 US$$_{1997}$$ per installed kW_e. Quoted from Solantausta et al. 1996 by Oonk et al. 1997.

[19] Scaled on gas turbine size.

TABLE 2.5 (CONTINUED)
Costs of System Components in MUS$$_{2001}$[1]

Unit	Base Investment Cost (fob)	Scale Factor	Base Scale	Overall Installation Factor[22]	Maximum Size[23]

[20] Steam system consists of water and steam system, steam turbine, condenser and cooling. Scaled on steam turbine size.

[21] Expansion turbine costs are assumed to be the same as steam turbine costs (without steam system).

[22] Overall installation factor. Includes auxiliary equipment and installation labor, engineering and contingencies. Unless other values are given by literature, the overall installation factor is set 1.86 for a 70-MW$_{th}$ scale (Faaij et al. 1998). This value is based on 33% added investment to hardware costs (instrumentation and control 5%, buildings 1.5%, grid connections 5%, site preparation 0.5%, civil works 10%, electronics 7%, and piping 4%) and 40% added installation costs to investment (engineering 5%, building interest 10%, project contingency 10%, fees/overheads/profits 10%, and start-up costs 5%). For larger scales, the added investments to hardware decreases slightly.

[23] Maximum sizes from Tijmensen (2000).

uncertainty range, one should be careful in ranking the concepts. Methanol 4 and 6 perform somewhat better than the other concepts due to an advantageous combination of lower investment costs and higher efficiency. The lowest methanol production price is found for concepts using the BCL gasifier, having lower investment costs. The combination of an expensive oxygen fired-IGT gasifier

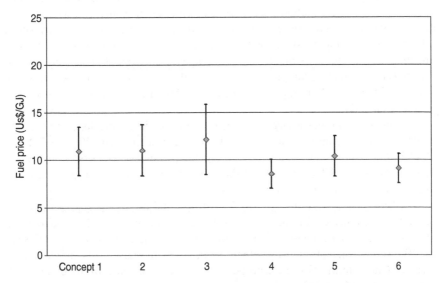

FIGURE 2.7 Methanol price for 400 MW$_{th}$ input concepts, with 30% uncertainty on investment and O&M (because O&M is a linear function of investment).

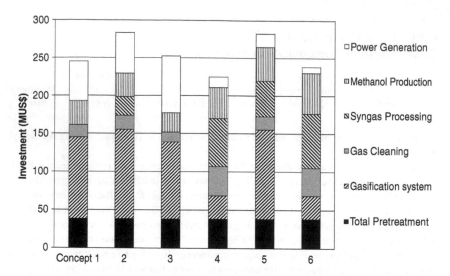

FIGURE 2.8 Breakdown of investment costs for 400 MW$_{th}$ concepts.

with a combined cycle seems generally unfavorable, since the efficiency gain is small compared to the high investment.

Investment redemption accounts for 42–76% of the annual costs and is influenced by the unit investment costs, the interest rate and the plant scale. The build-up of the total investment for all concepts is depicted in Figure 2.8. It can be seen that the costs for the gasification system (including oxygen production), synthesis gas processing and power generation generally make up the larger part of the investment. For autothermal reforming higher investment costs (Oonk et al. 1997) would increase the methanol price from considered concepts by about 1.5 US$/GJ. Developments in gasification and reforming technology are important to decrease the investments. On the longer term, capital costs may reduce due to technological learning: a combination of lower specific component costs and overall learning. A third plant built may be 15% cheaper leading to an 8–15% product cost reduction.

The interest rate has a large influence on the methanol production costs. At a rate of 5% methanol production costs decrease with about 20% to 7.2–9.0 US$/GJ. At a high-interest rate (15%), methanol production costs become 9.9–14 US$/GJ. Going to 1000 and 2000 MW$_{th}$ scales, the methanol production costs reach cost levels as low as 7.1–9.5 US$/GJ.

Feedstock costs account for 36–62% of the final product costs for the mentioned technologies. If a biomass price of 1.7 US$/GJ could be realized (a realistic price for, e.g., Brazil), methanol production costs would become 8.0–11 US$/GJ for 400 MW$_{th}$ concepts. On the other hand, when biomass costs increase to 3 US$/GJ (short term Western Europe), the production cost of methanol will increase to 10–16 US$/GJ.

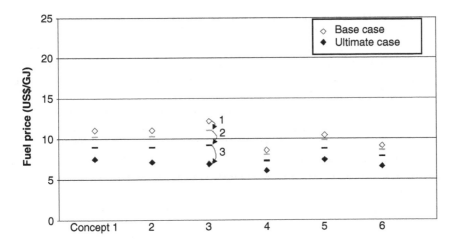

FIGURE 2.9 Optimistic view scenario. Different cost reductions are foreseeable: (1) biomass costs 1.7 US$/GJ instead of 2 US$/GJ, (2) technological learning reduces capital investment by 15% and (3) application of large scale (2000 MW$_{th}$) reduces unit investment costs.

If the electricity can be sold as green power, including a *carbon neutral* premium, the fuel production costs for power coproducing concepts drops, where the green premium essentially pays a large part of the fuel production costs. A power price of 0.08 US$/GJ would decrease methanol costs to –0.6–9.5 US$/GJ. Of course the decrease is the strongest for concepts producing more electricity. A green electricity scenario, however, may be a typical western European scenario. As such it is unlikely that it can be realized concurrent with biomass available at 1.7 US$/GJ.

In the long term, different cost reductions are possible concurrently (Tijmensen 2000). Biomass could be widely available at 1.7 US$/GJ, capital costs for a third plant built are 15% lower, and the large (2000 MW$_{th}$) plants profit from economies of scale. These reductions are depicted in Figure 2.9: methanol concepts produce between 6.1–7.4 US$/GJ. The influence of capital redemption on the annual costs has strongly reduced and the fuel costs of the different concepts lie closer together.

Previous analyses on short-term methanol production by Katofsky (1993) and Williams et al. (372 MW$_{HHV}$, 3.4 US$/GJ$_{HHV}$ feedstock, 0.07 US$/kWh$_e$ (Williams 1995; Williams et al. 1995)) yielded similar energy efficiencies (54–61% by HHV), but significantly higher methanol production costs: 14–17 US$/GJ$_{HHV}$. The largest difference is in the higher capital costs: higher TCI and higher annuity give 25–50% higher annual capital costs. The ADL/GAVE study (Arthur D. Little 1999) reports 13 US$/GJ methanol (feed 2 US$/GJ, 433 MW input) largely using input parameters from Katofsky. Komiyama et al. (2001) instead give much lower costs than presented here: 5 US$/GJ$_{HHV}$ for methanol at 530 MW$_{HHV}$ biomass input. However, in that study, process efficiencies and biomass cost are not given and a significant amount of energy is added as LPG.

In these long-term scenarios, methanol produced from biomass costs considerably less than methanol at the current market, which is priced about 10 US$/GJ (Methanex 2001). For application as automotive fuel, comparison with gasoline and diesel is relevant. Their production costs vary strongly depending on crude oil prices, but for an indication: 2003 gasoline market prices were about 7 US$/GJ at oil prices of 25–30 US$/bbl (BP 2004). DOE/EIA projects the world oil price in 2013 to amount between 17 and 34 US$, crude oil prices may decline as new deepwater oil fields are brought into production in the Gulf of Mexico and West Africa, new oil sands production is initiated in Canada, and OPEC and Russia expand production capacity (DOE/EIA 2005). In 2004 the average oil price was some 35 US$/bbl and currently even higher prices of about 50 US$/bbl are paid.

CONCLUSIONS

Methanol can be produced from wood via gasification. Technically, all necessary reactors exist and the feasibility of the process has been proven in practice. Many configurations are possible, of which several have been discussed in this chapter. The configurations incorporated improved or new technologies for gas processing and synthesis and were selected on potential low cost or high-energy efficiency. Some configurations explicitly coproduced power to exploit the high efficiencies of once-through conversion. The overall HHV energy efficiencies remain around 55%. Accounting for the lower energy quality of fuel compared to electricity, once-through concepts perform better than the concepts aiming at fuel only production. Also hot gas cleaning generally shows a better performance. Some of the technologies considered in this chapter are not yet fully proven/commercially available. Several units may be realized with higher efficiencies than considered here. For example, new catalysts and carrier liquids could improve liquid-phase methanol single-pass efficiency. At larger scales, conversion and power systems (especially the combined cycle) may have higher efficiencies, but this has not been researched in depth.

The methanol production costs are calculated by dividing the total annual costs of a system by the produced amount of methanol. Unit sizes, resulting from the plant modelling, are used to calculate the total installed capital of methanol plants; larger units benefit from cost advantages. Assuming biomass is available at 2 US$/GJ, a 400 MW_{th} input system can produce methanol at 9–12 US$/GJ, slightly above the current production from natural gas prices. The outcomes for the various system types are rather comparable, although concepts focussing on optimized fuel production with little or no electricity coproduction perform somewhat better.

The methanol production cost consists of about 50% of capital redemption, of which the bulk is in the gasification and oxygen system, synthesis gas processing and power generation units. Further work should give more insight into investment costs for these units and their dependence to scale. The maximum possible scale of particularly the pressurized gasifier gives rise to discussion. The operation and maintenance costs are taken as a percentage of the total investment, but may depend on plant complexity as well. Long-term (2020) cost reductions mainly reside in slightly lower biomass costs, technological learning, and application of large scales (2000 MW$_{th}$). This could bring the methanol production costs to about 7 US\$/GJ, which is in the range of gasoline/diesel.

Methanol from biomass could become a major alternative for the transport sector in a world constrained by greenhouse gas emission limits and high oil prices.

REFERENCES

Agterberg, A. *Bio-energy trade: Possibilities and constraints on short and longer term*, Utrecht University, Department of Science, Technology and Society, The Netherlands, 1998.

Alderliesten, P.T., *System study on high temperature gas cleaning, Part 2.3: Alkalimetals and other trace elements (in Dutch: Systeemtstudie hoge-temperatuur gasreiniging —deelstudie 2.3: Alkalimetalen en overige spore-elementen)*, Energy Research Center of the Netherlands ECN, Petten, 1990.

Armor, J.N., Applications of catalytic inorganic membrane reactors to refinery products, *Journal of Membrane Science*, 147, 217–233, 1998.

Little, A.D., *Analysis and integral evaluation of potential CO2-neutral fuel chains, GAVE reports 9908 - 9910 (Management Summary, Sheetpresentation and Appendices)*, Netherlands Agency for Energy and the Environment Novem, Utrecht, 1999.

Bechtel, Baseline design/economics for advanced Fischer-Tropsch technology, Topical report VI Natural gas Fischer-Tropsch case, Volume II Plant design and Aspen process simulation model, Bechtel, Pittsburgh, 1996.

Bensaïd, B., *Panorama 2005: Alternative Motor Fuels Today and Tomorrow*, Institute Francais du Petrol—IFP, Rueil-Malmaison France, 2004.

Bergman, P.C.A., van Paasen, S.V.B. and Boerrigter, H., The novel "OLGA" technology for complete tar removal from biomass producer gas, In *Pyrolysis and Gasification of Biomass and Waste*, Bridgwater, A.V., Ed., CPL Press, Newbury, U.K., 2003 pp. 347–356.

Boerrigter, H., den Uil, H. and Calis, H., Green diesel from biomass via Fischer-Tropsch synthesis: New insights in gas cleaning and process design, In *Pyrolysis and Gasification of Biomass and Waste*, Bridgwater, A.V., Ed., CPL Press, Newbury, U.K., 2003, pp. 385-394.

BP, 2004 *Statistical Review of World Energy*, BP, London, U.K., 2004, www.bp.com.

Brown, F.C., Alternative uses for methanol plants, Jacobs Engineering Ltd., London, 1999, http://www.jacobsconsultancy.com/pdfs/petech.pdf.

Consonni, S. and Larson, E., Biomass-gasifier/aeroderivative gas turbine combined cycles, Part A: Technologies and performance modeling, Part B: Performance calculations and economic assessment, In *The American Society of Mechanical Engineers' 8th Congress on Gas Turbines in Cogeneration and Utility, Industrial and Independent Power Generation*, ASME/IGTI, Portland, OR, 1994.

Cybulski, A., Liquid Phase Methanol Synthesis: Catalysts, Mechanism, Kinetics, Chemical Equilibria, Vapor-Liquid Equilibria, and Modeling—A Review, Catalytic Review—*Science Engineering*, 36 (4), 557–615, 1994.

de Lathouder, H.C., Process for the preparation of methanol, European patent specification EP0111376, 1982.

DeLallo, M.R., Rutkowski, M. and Temchin, J., *Decarbonized fuel production facility— a technical strategy for coal in the next century, Article 00844,* Parsons Power Group, Inc., Reading, PA, 2000.

DOE/EIA, Annual Energy Outlook 2005, U.S. Department of Energy, Energy Information Administration, Washington, DC, 2005.

Dybkjaer, I. and Hansen, J.B., Large-scale production of alterantive synthetic fuels from natural gas, *Studies in Surface Science and Catalysis*, 107, 99–116, 1997.

Ekbom, T., Lindblom, M., Berglin, N. and Ahlvik, P., *Technical and Commercial Feasibility Study of Black Liquor Gasification with Methanol/DME Production as Motor Fuels for Automotive Uses—BLGMF, EU Altener Contract 4.1030/Z/01-087/2001,* Nykomb Synergetics AB, Stockholm, 2003.

Faaij, A., Meuleman, B. and van Ree, R., *Long term perspectives of biomass integrated gasification with combined cycle technology,* Netherlands Agency for Energy and the Environment Novem, Utrecht, 1998.

Faaij, A., Ree, R.V. and Oudhuis, A., *Gasification of biomass wastes and residues for electricity production: Technical, economic and environmental aspects of the BIG/CC option for the Netherlands,* Utrecht University, Department of Science, Technology and Society, The Netherlands, 1995.

Faaij, A.P.C., Energy from biomass and waste, Ph.D. thesis, Utrecht University, Department of Science, Technology and Society, The Netherlands, 1997.

Farla, J.C.M., Hendriks, C.A. and Blok, K., Carbon dioxide recovery from industrial processes, *Climatic Change*, 29, 439–461, 1995.

Gas Turbine World, Gas Turbine World 1997 Handbook; Pequot Publishing, Fairfield, Canada, 1997.

Hagihara, K., Mabuse, H., Watanabe, T., Kawai, M. and Saito, M., Effective liquid-phase methanol synthesis utilizing liquid-liquid separation, *Energy Conversion and Management*, 36 (6–9), 581–584, 1995.

Haldor Topsoe, Hydrogen Production by Steam Reforming of Hydrogen Feedstocks, Haldor Topsoe, 1991.

Hall, D.O., Rosillo-Calle, F., Williams, R.H. and Woods, J., Biomass for energy: supply prospects, In *Renewable Energy Sources for Fuels and Electricity*, Johansson, T.B., Kelly, H., Reddy, A.K.N. and Williams, R.H., Eds., Island Press, Washington, DC, 1992, pp. 593–651.

Hamelinck, C.N. and Faaij, A.P.C., *Future prospects for production of methanol and hydrogen from biomass*; Utrecht University, Copernicus Institute, Science Technology and Society, The Netherlands, 2001.

Hendriks, C., Carbon dioxide removal from coal-fired power plants, Ph.D. thesis, Department of Science, Technology and Society, Utrecht University, The Netherlands, 1994.

Hepola, J. and Simell, P., Sulphur poisoning of nickel-based hog gas cleaning catalysts in synthetic gasification gas, II. Chemisorption of hydrogen sulphide, *Applied Catalysis B: Environmental*, 14, 305–321, 1997.

Heydorn, E.C., Diamond, B.W. and Lilly, R.D., *Commercial-scale demonstration of the liquid phase methanol (LPMEOHTM) process*, Air Products and Chemicals, Inc. and Eastman Chemical Company, Allentown, PA, 2003.

Hydrocarbon Processing, *Gas Processes '98*, 1998.

Ishibashi, M., Otake, K. and Kanamori, S., Study on CO_2 removal technology from flue gas of thermal power plant by physical adsorption method, In *Fourth International Conference on Greenhouse Gas Control Technologies*, Interlaken, 1998.

Jansen, D., *System study on high temperature gas cleaning, Part 2.1: H2S/COS removal (in Dutch: Systeemstudie hoge-temperatuur gasreiniging - deelstudie 2.1: H2S/COS-verwijdering)*, Energy Research Center of the Netherlands ECN, Petten, 1990.

Jothimurugesan, K., Adeyiga, A.A. and Gangwal, S.K., Advanced hot-gas desulfurization sorbents, In *Advanced Coal-Fired Power Systems Review Meeting*, Morgantown, WV, 1996

Katofsky, R.E., *The production of fluid fuels from biomass*, Princeton University, Center for Energy and Environmental Studies, NJ, 1993.

Kirk-Othmer, *Encyclopedia of Chemical Technology, 4th ed.*, 1995.

Klein Teeselink, H. and Alderliesten, P.T., *System study on high temperature gas cleaning, Part 2.4: removal of particles (in Dutch: Systeemtstudie hoge-temperatuur gas-reiniging - deelstudie 2.4: stofverwijdering)*, Stork Ketels, B.V., Hengelo, 1990.

Knight, R. Personal communications on the pressurized renugas gasifier for different conditions, Institute of Gas Technology, Chicago, 1998.

Komiyama, H., Mitsumori, T., Yamaji, K. and Yamada, K., Assessment of energy systems by using biomass plantation, *Fuel*, 80, 707–715, 2001.

Kurkela, E., *Formation and removal of biomass-derived contaminants in fluidized-bed gasification processes*, VTT Offsetpaino, Espoo, Finland, 1996.

Kwant, K.W., *Status of gasification in countries participating in the IEA Bioenergy gas-ification activity*, Netherlands Agency for Energy and the Environment, Novem, Utrecht, 2001.

Larson, E., Consonni, S. and Kreutz, T., Preliminary economics of black liquor gasifier/gas turbine cogeneration at pulp and paper mills, In *The 43rd ASME Gas Turbine and Aeroengine Congress, Exposition and Users Symposium*, Stockholm, 1998.

Larson, E.D. and Marrison, C.I., Economic scales for first-generation biomass-gasifier/gas turbine combined cycles fueled from energy plantations, *Journal of Engineering for Gas Turbines and Power*, 119, 285–290, 1997.

Li, X., Grace, J.R., Watkinson, A.P., Lim, C.J. and Ergüdenler, A., Equilibrium modeling of gasification: a free energy minimization approach and its application to a circulating fluidized bed coal gasifier, *Fuel*, 80 (2), 195–207, 2001.

Maiya, P.S., Anderson, T.J., Mieville, R.L., Dusek, J.T., Picciolo, J.J. and Balachandran, U., Maximizing H2 production by combined partial oxidation of CH4 and water gas shift reaction, *Applied Catalysis A: General*, 196, 65–72, 2000.

Methanex, http://www.methanex.com, 2001.

Methanol Institute, Factsheet: World methanol supply/demand, Methanol Institute, Arling-ton, VA, 2005.

Milne, T.A., Abatzoglou, N. and Evans, R.J., Biomass gasifier 'tars': their nature, formation and conversion, National Renewable Energy Laboratory, report NREL/TP-570-25357, 1998.

Mitchell, S.C., *Hot gas particulate filtration*, International Energy Agency IEA Coal Research, London, 1997.

Mitchell, S.C., *Hot gas cleanup of sulphur, nitrogen, minor and trace elements*, International Energy Agency IEA Coal Research, London, 1998.

OECD, *National accounts—Main aggregates, Edition 1997*, Paris, 1996.

Oonk, H., Vis, J., Worrell, E., Faaij, A. and Bode, J.-W., *The MethaHydro-process—preliminary design and cost evaluation*, TNO, The Hague, 1997.

OPPA, Assesment of costs and benefits fo flexible and alternative fuel use in the US transportaion sector, Technical Report 5: Costs of methanol production from biomass, U.S. Department of Energy, Washington, DC, 1990.

Paisley, M.A., Farris, M.C., Black, J., Irving, J.M. and Overend, R.P., Commercial demonstration of the Battelle/FERCO Biomass gasification process: startup and initial operating experience, In *Fourth Biomass Conference of the Americas*, Overend, R.P., Chornet, E., Eds., Elsevier Science, Oxford, Oakland, CA, 1998, pp. 1061–1066.

Peckham, J., Lurgi: cheaper gtl diesel route through methanol, *Gas-to-Liquids News*, 6 (11), http://www.worldfuels.com., 2003.

Perry, R.H., Green, D.W. and Maloney, J.O., *Perry's Chemical Engineers' Handbook, Sixth Edition*, McGraw-Hill Book Co., Singapore, 1987.

Peters, M.S. and Timmerhaus, K.D., *Plant Design and Economics for Chemical Engineers, Third Edition*, McGraw-Hill Book Co., New York, 1980.

Pierik, J.T.G. and Curvers, A.P.W.M., *Logistics and pretreatment of biomass fuels for gasification and combustion*, Energy Research Center of the Netherlands ECN, Petten, 1995.

Pieterman, M. The historical development of innovative and energy-efficient synthesis gas production technologies (M.S. thesis), Utrecht University, Department of Science, Technology and Society, The Netherlands, 2001.

Riesenfeld, F.C. and Kohl, A.L., *Gas Purification*; Gulf Publishing Company, Houston, 1974.

Rodrigues de Souza, M., Walter, A. and Faaij, A., An analysis of scale effects on co-fired BIG-CC system (biomass + natural gas) in the state of São Paulo/Brazil, In *1st World Conference on Biomass for Energy and Industry*, Kyritsis, S., Beenackers, A.A.C.M., Helm, P., Grassi, A. and Chiaramonti, D., Eds., James & James, London, Sevilla, 2000, pp. 813–816.

Solantausta, Y., Bridgewater, T. and Beckman, D., *Electricity production by advanced biomass power systems*, Technical Research Centre of Finland (VTT), Espoo, Finland, 1996.

Suda, T., Fujii, M., Yoshida, K., Iijima, M., Seto, T. and Mitsuoka, S., Development of flue gas carbon dioxide recovery technology, In *First International Conference on Carbon Dioxide Removal*, Blok, K., Turkenburg, W.C., Hendriks, C.A. and Steinberg, M., Eds., Pergamon Press, Amsterdam, 1992.

Tijm, P.J.A., Brown, W.R., Heydorn, E.C. and Moore, R.B., *Advances in Liquid Phase Technology—Presentation at American Chemical Society Meeting, April 13–17*, San Francisco, 1997.

Tijmensen, M.J.A., The production of Fischer Tropsch liquids and power through biomass gasification (M.S. thesis), Utrecht University, Department of Science, Technology and Society, The Netherlands, 2000.

Turn, S.Q., Kinoshita, C.M., Ishimura, D.M., Zhou, J., Hiraki, T.T. and Masutani, S.M., A review of sorbent materials for fixed bed alkali getter systems in biomass gasfier combined cycle power generation applications, *Journal of the Institute of Energy*, 71, 163–177, 1998.

U.S. Department of Energy, *Commercial-scale demonstration of the liquid phase methanol (LPMEOH TM) process—Clean Coal Technology Topical Report #11*; U.S. Department of Energy, 1999.

van den Hoed, R., Driving fuel cell vehicles—How established industries react to radical technologies (Ph.D. thesis), Delft University of Technology, The Netherlands, 2004.

van Dijk, K.M., van Dijk, R., van Eekhout, V.J.L., van Hulst, H., Schipper, W. and Stam, J.H., *Methanol from natural gas—conceptual design & comparison of processes*, Delft University of Technology, The Netherlands, 1995.

van Ree, R., *Air separation technologies—An inventory of technologies for 'pure' oxygen production for pulverised coal combustion in a CO2(g)/O2(g)-atmosphere (in Dutch)*; Energy Research Center of the Netherlands ECN, Petten, 1992.

van Ree, R., Oudhuis, A., Faaij, A. and Curvers, A., *Modelling of a biomass integrated gasifier/combined cycle (BIG/CC) system with the flowsheet simulation programme ASPEN+*, Energy Research Center of the Netherlands ECN and Utrecht University, Department of Science, Technology and Society, Petten, 1995.

van Walwijk, M., Bückmann, M., Troelstra, W.P. and Achten, P.A.J., *Automotive fuels survey—Part 2 Distribution and use*, IEA/AFIS operated by Innas bv, Breda, the Netherlands, 1996.

Verschoor, M.J.E. and Melman, A.G., *System study high temperature gas cleaning at IGCC systems*, NOVEM / TNO Environment Energy and Process innovation MEP, 1991.

White, L.R., Tompkins, T.L., Hsieh, K.C. and Johnson, D.D., Ceramic filters for hot gas cleanup, In *International Gas Turbine and Aeroengine Congress and Exposition*, Cologne, Germany, 1992, p. 8.

Williams, R.H., Larson, E.D., Katofsky, R.E., *J.* Chen methanol and hydrogen from biomassa for transportation. *Energy for Sustainable Development*, I (5), 18–34, 1995.

Williams, R.H., *Cost-competitive electricity from coal with near-zero pollutant and CO2 emissions—review draft*, Princeton University, Center for Energy and Environmental Studies, NJ, 1998.

Williams, R.H., Larson, E.D., Katofsky, R.E. and Chen, J. *Methanol and hydrogen from biomass for transportation, with comparisons to methanol and hydrogen from natural gas and coal, PU/CEES Report 292*, Princeton University, Center for Energy and Environmental Studies, NJ, 1995.

Wilson, M.A., Wrubleski, R.M. and Yarborough, L., Recovery of CO2 from power plant flue gases using amines, In *First International Conference on Carbon Dioxide Removal*, Blok, K., Turkenburg, W.C., Hendriks, C.A. and Steinberg, M., Eds., Pergamon Press, Amsterdam, 1992, pp. 325–331.

Yakobson, D.L., Fischer-Tropsch Technology: New Project Opportunities (Rentech, Inc.), In *Gas-To-Liquids Processing 99 Meeting*, Intertech Corporation, Portland, ME, 1999.

3 Landfill Gas to Methanol

William H. Wisbrock
President, Biofuels of Missouri, Inc., St. Louis

CONTENTS

Abstract An explanation of how market forces, governmental mandates, and tax incentives have placed the use of landfill gas as an alternative energy source into a growing industry in the United States. Also included is a description of the opportunities and challenges that face this emerging domestic energy industry, along with a description of the Acrion Technology CO_2 Wash process that cleans landfill gas, so that it can be utilized for the manufacture of methanol.

LANDFILLS AND LANDFILL GAS

Landfills are physical facilities used for the disposal of residual solid wastes in the surface soils of the earth. Historically, landfills have been the most economical and environmentally acceptable method for the disposal of solid wastes, both in the United States and throughout the world. Even with the implementation of waste reduction, recycling, and transformation technologies, nearly all of the residual solid waste in the United States today is deposited in landfills. Furthermore, landfills are not going to disappear, rather they will continue to be an important component of solid waste management strategy far into the twenty-first century.

Landfills produce a large amount of gas. Anaerobic decomposition of the biodegradable portion of the municipal solid waste produces methane and carbon dioxide in roughly equal amounts. These two principal components, together with atmospheric nitrogen and oxygen and trace organic compounds, comprise landfill gas, LFG. According to the Environmental Protection Agency (EPA) Landfill

51

Methanol Outreach Program (LMOP) statistics, each pound of organic waste biodegrades into 10 to 12 standard cubic feet of gas during its landfill residence of approximately 25 years. Modest size landfills produce one to five million standard cubic feet of landfill gas daily. By way of example, one of the largest landfills in the United States, Fresh Kills, Staten Island, NY, produces more than 30 million cubic feet of landfill gas daily. Landfill gas generation increases while the landfill is active and decays three to five percent annually beginning several years after the landfill is closed. Significant landfill gas is generated for up to 25 years after closure of the landfill.

The 1986 Clean Air Act (CAA) requires that landfills containing over 2.5 million tons of municipal solid waste be required to collect and flare the landfill gas in order to prevent methane migration and control the odor associated with the landfill. This requirement helps prevent methane migration, which contributes to local smog and global climate change. Methane will try and escape into the atmosphere from the landfill either through fissures in the lining of the landfill or through the surface cover. Landfills are the largest human-related source of methane in the United States, accounting for about 34% of all methane emissions. The amount of methane created depends on the quantity and moisture content of the waste and the design and management and environmental practices of the landfill.

In order to comply with the CAA, LFG is extracted from the landfill by an engineered system of liners, pipes, wells, wellhead monitors, and a vacuum system to move the collected gas to a metering device and then to the constant temperature flare. Significant amounts of landfill gas that are now flared could be utilized for economically viable projects. The EPA estimates that there are between 800 and 1000 domestic landfills that are currently flaring landfill gas that could be converted to energy and energy-related projects, thereby reducing dependence on fossil energy. Methane vented or flared from existing U.S. landfills was estimated by the LMOP in 2001 to equal about 5% of domestic natural gas consumption or about 1% of domestic total energy needs.

Landfill gas is similar to low-quality natural gas in that it requires the removal of the volatile organic contaminants and the CO_2 to realize substantial commercial value. Landfill gas contaminants challenge separation technology because the potential contaminants can number in the hundreds of chemical compounds and various toxic species such as vinyl chloride and hydrogen sulfide. Additionally, no two landfills have the same contaminants and these contaminants are constantly changing over the gas production life of the landfill as the decomposition occurs.

The conventional uses of landfill gas to energy include electricity generation using internal combustion engines, turbines, micro turbines and fuel cells; direct use, which would include boiler, dryer, kiln, greenhouse, wastewater treatment; cogeneration, also known as combined heat, and power that enjoys the efficiency of capturing the thermal energy in addition to electricity generation; and alternative fuels that include pipeline quality gas, compressed natural gas, liquefied natural gas, methanol, and hydrogen.

Landfill ownership is either public or private. Solid waste disposal firms generally own and operate the majority of the private landfills. Privately owned landfills tend to promote their gas resource and solicit buyers or users of the gas more aggressively than their public counterparts. The process of obtaining and acquiring landfill gas rights is essentially the same for either case. Such a process usually consists of the following: a review of the proposal by the owner or appropriate public officials, the negotiation of the business plan and definition of responsibilities and liabilities, and an execution of a contractual agreement governing the gas rights, responsibilities of the parties, term of the agreement, price for the gas, and other details of mutual concern.

Establishing a price for landfill gas and other project considerations requires the examination and negotiation of many factors, including but not limited to the following:

- Amount of landfill gas available and the projections of future gas generation rates.
- Gas composition or gas quality.
- Environmental regulations and permits required.
- Ownership of gas collection system and responsibility for its maintenance.
- Competing prices for natural gas and electric in the area.
- IRS Section 29 tax credit availability.
- Building permits and access to landfill.
- Local air quality conditions and regulations.

The IRS Section 29 tax credits were an attempt to provide a financial incentive for the utilization of the landfill gas for energy projects. The IRS Code provided for a $1.05 per million BTU tax credit to a landfill gas developer if such an energy project were started prior to June 30, 1998. These tax credits will expire on December 31, 2007, and no longer provide financial incentive to promote traditional landfill gas to energy projects.

METHANOL, PRESENT, AND FUTURE

Methanol (chemical formula CH_3OH and also known as methyl alcohol or wood alcohol) is a clear, colorless liquid that is water soluble and readily biodegradable. Methanol occurs naturally in the atmosphere as a by-product of biomass and landfill decomposition. As an industrial product, methanol is a liquid petrochemical that can be made from renewable and nonrenewable fossil fuels containing carbon and hydrogen. Since natural gas costs account for the major portion of the operating costs of domestic methanol producers and are followed in importance by distribution costs and operating costs, virtually all new methanol production has been moved offshore near low cost or "stranded" natural gas locations.

Large world-scale or megamethanol plants are being built in these stranded gas areas, such as Chile, Trinidad, Qatar, Equatorial Guinea, and Saudi Arabia.

Each plant can produce 300 million gallons annually of methanol and costs more than $350 million to construct. The construction of these megaplants has reinforced the characterization of methanol as a "typical" commodity as cycles of oversupply resulting in lower prices and idled capacity are followed by periods of shortage and rapidly rising prices as demand catches up and exceeds supply until increased prices justify new plant investment.

Prior to the 1980s, methanol was produced and consumed locally in North America primarily as an intermediate feedstock for derivatives such as formaldehyde, acetic acid, and plastics recycling in packaging. Limited international trade was seen mostly through U.S. exports. U.S. natural gas feedstock, at that time, was reasonably competitive. Today methanol is a global commodity and the United States has gone from the position of largest producer in the world, to the largest net importer in less than a decade.

The American Methanol Institute, the trade organization for the methanol industry, has changed its name to the Methanol Institute and predicts that there will be no U.S.-produced methanol within five years due primarily to high and volatile natural gas prices. With the controversy surrounding methyl-tertiary-butyl-ether (MTBE), significant demand for methanol has disappeared from the domestic market. While it was widely predicted that methanol prices over the past several years would be extremely weak, the recovery of the international economy and demand from China and India have virtually made up for the rapid decline for the demand of methanol for MTBE.

MTBE was developed in the early 1990s as an oxygenate to improve more complete combustion of gasoline. This was done at the behest of the EPA. However, leaking underground storage tanks at automobile service stations caused ground water contamination and MTBE was found to be a possible carcinogen. As a result, California, New York, and several other states have banned MTBE. It has largely been replaced by ethanol as an oxygenate for gasoline.

LANDFILL GAS TO METHANOL

Most existing and proposed landfill gas recovery projects simply combust raw or minimally treated landfill gas to generate heat or electricity. No competitive technology for landfill gas recovery has yet to capture any significant market share of the available landfill gas resources. Landfill gas to methanol is technically the most challenging issue since contaminant removal to parts per billion is required to manufacture methanol and made possible only by the CO_2 Wash™ technology developed and patented by Acrion Technology, Inc. of Cleveland, OH. The CO_2 Wash™ process is shown in Figure 3.1.

CO_2 Wash™ removes the contaminants from landfill gas using liquid carbon dioxide condensed directly from the landfill gas. A stream of contaminant-free methane and carbon dioxide is produced, along with a condensed stream of contaminants in carbon dioxide. This intermediate stream of clean methane and carbon dioxide can be used as fuel gas or as feedstock to make methanol. This second process is shown in Figure 3.2. In the alternative, this intermediate stream

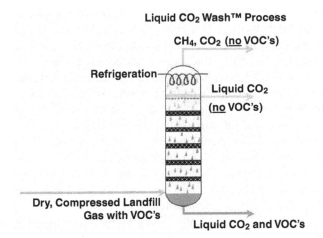

FIGURE 3.1 The CO$_2$ wash process.

can be further processed to separate the carbon dioxide from the methane to produce methane as pipeline gas or transportation fuel (compressed or liquid) and liquid carbon dioxide. The contaminant-laden carbon dioxide stream is incinerated in the landfill flare in all cases as is being done if there were no LFG to energy project.

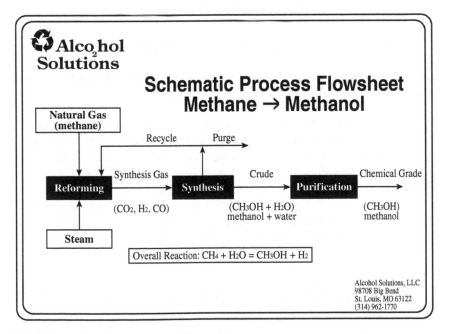

FIGURE 3.2 The methane-to-methanol process.

Landfill gas is pretreated to remove water and hydrogen sulfides and then compressed to about 400 psi. Dry, compressed landfill gas is introduced to a conventional packed tower where the contaminants are removed. Pure liquid carbon dioxide condenses from the treated landfill gas at the top of the column. A small portion of the liquid carbon dioxide is sent down the column (descending drops) where, in intimate contact with rising landfill gas, it absorbs (or washes) contaminants from the landfill gas. The larger portion of this condensed liquid is available as liquid carbon dioxide product at the top of the column. Clean landfill gas feedstock for methanol synthesis is at the top of the column, with a composition of 75% methane, 25% carbon dioxide.

In 1998, Acrion completed successful field tests of the CO_2 Wash at the Al Turi Landfill in Goshen, NY, under a grant from the U.S. Department of Energy (DOE). In the fall of 2001, a demonstration CO_2 Wash™ skid, also funded by a U.S. DOE grant, was operated successfully at the New Jersey EcoComplex, Burlington County, NJ. Continuity and quality of products were established in an extended 100+-hour run producing over 3500 lbs of liquid carbon dioxide (>0.9999 purity). Atlantic Analytical Laboratory and the EPA approved the facility for landfill gas testing and confirmed the product gas streams purity. Mack Truck has verified technology veracity in 2003–2004 trials and is proceeding with additional liquid natural gas production programs.

This CO_2 Wash™ process experience, combined with the extensive experience that HydroChem, a subsidiary of Linde, AG, has had in building modular hydrogen/methanol plants, has reduced significantly the investment and business concerns associated with landfill gas to methanol projects.

RENEWABLE METHANOL

Utilizing landfill gas to manufacture methanol has many significant advantages over the current suppliers of methanol.

- A known price for the raw material and the known price of the plant allows the producer to control the production costs of methanol over the productive life of the landfill, typically the remaining years that the landfill will be open and then 25 years after it is closed.
- The ability to determine production costs enables the manufacturer to offer long-term, price-stable contracts to customers who have traditionally experienced wide fluctuations in the price of methanol.
- The proximity of the production site to the methanol customers results in considerably lower transportation costs and more dependable delivery.
- Being classified as a renewable resource, the manufacturer of methanol from landfill gas enjoys a "green" reputation in the United States. A number of Fortune 500 companies have a corporate policy to purchase a certain percentage of their raw material needs from renewable sources.

- "Green methanol" will allow the manufacturer to take advantage of various tax credits and incentives passed by the United States Congress to promote renewable products and fuels. One example is the 60-cents-per-gallon "alcohol fuel credit" that applies to methanol used in transportation fuels.
- The utilization of landfill gas in the manufacture of methanol results in the sequestering of methane and carbon dioxide, this process is credited with reducing greenhouse gas emissions.

The emerging markets for renewable methanol are becoming more defined and mature as demand is being driven by environmental statutory requirements and U.S. government mandates regarding the reduction of global warming gases and ending U.S. reliance upon foreign oil and energy. To that end, landfill gas to methanol offers a compelling opportunity to develop a significantly large domestic industry to support the growing demand for domestic renewable fuels and energy.

NEW USES FOR DOMESTIC METHANOL

- Methanol fuel cells: transportation, stationary, and portable power.
- Production of biodiesel.
- Sewerage treatment denitrification.
- Fuel for standby turbine electric generators.
- Pulp and paper bleaching replacing chlorine.
- Stock-car racing fuel.
- Replacement for diesel fuel for stationary diesel engines.
- Converting methanol to hydrogen. Eliminating storage and transportation.
- Fuel additives for diesel fuel. Preventing gelling and fuel line freeze.
- Intermediate for production of dimethyl ether, acetic acid, and formaldehyde.
- Environmental cleanup of perchlorate at military installations.

All referenced data in this chapter comes from the U.S. Environmental Protection Agency (http://www.epa.gov).

4 The Corn Ethanol Industry

Nancy N. Nichols,[1] Bruce S. Dien,[1]
Rodney J. Bothast,[2] and Michael A. Cotta[1]

[1] Fermentation Biotechnology Research Unit, National Center for Agricultural Utilization Research, Agricultural Research Service, U.S. Department of Agriculture,* Peoria, Illinois

[2] National Corn-to-Ethanol Research Center, Southern Illinois University-Edwardsville, Edwardsville, Illinois

CONTENTS

---

* Mention of trade names or commercial products in this chapter is solely for the purpose of providing specific information and does not imply recommendation or endorsement by the U.S. Department of Agriculture.

INTRODUCTION

Ethanol production is a growing industry in the United States, where corn is the feedstock used to produce approximately 90% of fuel ethanol. Approximately 1.26 billion bushels of corn, equal to 11% of the total U.S. corn crop, were processed to ethanol in 2004 according to the Renewable Fuels Association. Globally, the only crop used to produce more ethanol is sugar cane. Approximately 61% of world ethanol production is from sugar crops, with the remainder being made primarily from corn [1]. The success of corn as a feedstock for ethanol production can be directly tied to the huge and sustained improvements in yields realized in the United States; corn yields per acre quadrupled in the fifty years from 1954 to 2004. According to the U.S. Department of Agriculture-National Agricultural Statistics Service, the average U.S. corn grain yield in 2004 was 160.4 bushels per acre and the average price was $1.95 per bushel.

Use of corn to make fuel ethanol dates back to the earliest days of automobiles [2]. The Model T was originally designed to run on ethanol, although use of ethanol for fuel ceased with the development of the petroleum industry. The ethanol industry was briefly revived during World War II and during the oil crisis in the 1970s. The current renaissance of the ethanol industry dates to the 1980s, when the U.S. Environmental Protection Agency mandated adding fuel oxygenates to gasoline to reduce automotive emissions. Ethanol has been used as an oxygenate primarily in the Midwestern United States whereas MTBE, which is derived from petroleum, has been used on both coasts. However, MTBE is now recognized as a hazard to water supplies, and its use is being phased out by all states. The opening of coastal markets to ethanol has led to tremendous growth in the domestic ethanol industry, which is expected to produce five billion gallons of ethanol per year by 2012. Whether this target of five or more billion gallons will be achieved because of oxygenate requirements, or be directly mandated as part of a National Energy Bill—which will allow gasoline refiners to trade ethanol credits—is unclear. What appears certain is that the corn ethanol industry can look forward to continued and steady growth. Growth is ensured by the increasingly appreciated advantages of corn ethanol: reduced oil imports, reduced automotive-associated net CO_2 emissions, and a stabilized corn market.

Corn Ethanol and the Energy Balance

Ethanol was promoted at the end of the twentieth century for its environmental benefits as an oxygenate in reducing organic carbon emissions. Now, however, the advantages associated with reduced CO_2 emissions and oil imports are increasingly cited. A prime concern in justifying the use of bioethanol is calculating the benefit of using it for fuel, from an energy savings basis. Most energy analyses have shown that ethanol contains more energy than the fossil fuels used to produce it [3–5]. The average energy output/input ratio is 1.57 and 1.77 for the wet-milling and dry-grind ethanol processes, respectively [6]. These values are life-cycle estimates, meaning that they include all energy inputs and outputs from growing the corn to transporting the ethanol to market. In other words, on average approximately 67% more energy captured from the sun (by photosynthesis) is retained in ethanol than the fossil fuel energy used to grow and harvest the corn and convert it to ethanol. By comparison, the energy obtained from gasoline is 20% lower than the fossil inputs for production [5].

Advances in agronomics, fertilizer production and application, and the ethanol production process have significantly decreased energy requirements, reducing by half the energy currently required for ethanol production compared to that required in the late 1970s. As the ethanol industry matures, the net energy gains are likely to increase further. A recent example is the cold-starch hydrolysis process, discussed later in this chapter, which uses less energy for ethanol production. Another key consideration regarding the use of fuel ethanol is how well it acts as a replacement for imported liquid transportation fuel. Much of the energy that goes into making ethanol is in the form of coal and natural gas used to operate fertilizer and ethanol plants. Therefore, it has been estimated that more than six gallons of ethanol are produced for each gallon of petroleum used [7].

Economics of Corn Ethanol

Advocates of the corn ethanol industry note that agricultural business and rural economies benefit from ethanol production. Rural economies gain new jobs and an expanded tax base. The local impact of a dry grind facility producing 40 million gallons of ethanol per year is an estimated $56 million spent annually for feedstock (corn), labor, and utilities, and also $1.2 million in state and local tax revenues [8]. Farmers benefit from the establishment of a guaranteed market for a set number of bushels, with an increased value of $0.25–$0.50 per bushel in the national corn price plus an additional $0.05–$0.10 per bushel in the locality of ethanol plants. A strengthened domestic market for corn also prevents the United States from flooding the global corn market.

Federal and state tax support for ethanol production has driven growth of the ethanol industry. Federal ethanol tax incentives for ethanol-blended fuels are directed at gasoline marketers in one of two ways. The excise tax exemption reduces the federal excise tax, paid at the terminal by refiners and marketers, by 5.1 cents per gallon of 10% ethanol-blended fuel. Or, gasoline refiners can claim an income tax credit of 5.1 cents per gallon of gasoline blended with 10% ethanol. At the state level, tax incentives generally are directed to benefit small ethanol producers, which are typically farmer-owned cooperatives [9]. In some cases, states directly support new ethanol facilities with cash payments to help defray construction costs. State incentives for users range from cents-per-gallon tax exemptions for ethanol blends to rebates for purchase of alternative fuel vehicles and grants to fuel stations selling alternative fuels. Several states mandate use of ethanol blends or flexible fuel vehicles for state-owned fleets. Minnesota and Hawaii have renewable fuel standards requiring use of E10 blends in cars, and Minnesota's E20 law offers two options for increasing ethanol use to 20 percent of the gasoline sold in the state by 2013. Montana's E10 law states that most of the gasoline sold in the state must include 10% ethanol when annual production of ethanol in the state reaches 40 million gallons. A few other state legislatures are considering similar measures.

THEORY BEHIND CONVERSION OF CORN TO ETHANOL

STRUCTURE AND COMPOSITION OF THE CORN KERNEL

Corn kernels contain, by weight, approximately 70% starch, 9% protein, 4% fat and oil, and 9% fiber on a dry basis [10]. Most corn grown for ethanol production is #2 yellow dent corn, so named because of the indentation in the top of the dried kernel. Energy is stored in the seed in the form of starch and oil, which are segregated to the endosperm and germ, respectively (Figure 4.1). Different proteins are contained in the endosperm, germ, and tip cap. The gluten protein fraction is found in the endosperm, bonded to the starch. The seed contents are protected by a waxy coat and fibrous outer layer (the pericarp). Fiber is also present in the germ and tip cap.

Ethanol yield potential varies among corn hybrids [11, 12], and also depends on agronomic practices and environmental factors. Corn hybrids are being developed and marketed specifically for enhanced ethanol production, and seed corn is labeled for sale with high extractable starch (for wet milling) or high fermentable starch content (for dry grind ethanol processing). Total starch content and total extractable starch content do not necessarily correlate with ethanol yields obtained in dry grind processing of the whole kernel; instead, high-performing varieties have been identified by empirical testing. Grain from high ethanol-

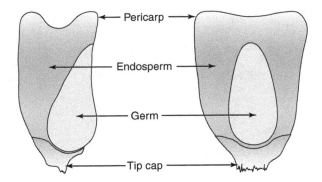

FIGURE 4.1 Corn kernel structure.

yielding hybrids reportedly results in ethanol yields up to 4% higher than the yield from mixed commercial grain, representing an additional one to two million dollars to a 40-million-gallon-per-year dry grind ethanol facility [13]. Spectroscopic methods using near-infrared (NIR) technology have been developed for use at ethanol plants to predict the ethanol yield potential of samples of whole corn kernels [14].

CONVERSION OF STARCH TO GLUCOSE

The wet milling and dry grind fermentation processes share the same biological basis for conversion of corn starch to ethanol: starch is converted by the combined actions of heat and enzymes to glucose and maltose, which are fermented by yeast to ethanol. Starch is a mixture of two glucose polymers: amylose, a linear molecule with α-1-4 linkages, and amylopectin, a branched molecule which has the same α-1-4 linkages and also contains α-1-6 branch points. Starch forms crystalline granules in the seed [15]. The granules (Figure 4.2) are insoluble in water and, in fact, have hydrophobic interiors. Pores extend from the surface into the hollow core of the granule. Heating an aqueous starch suspension weakens the hydrogen bonds within and between starch molecules, causing swelling of the starch granules due to absorption of water. The swelling process is called gelatinization [16]. Gelatinized starch is converted to glucose in the industrial process primarily by two enzymes, alpha-amylase and glucoamylase. First, the starch polymer is hydrolyzed by alpha-amylase to shorter chains called dextrins in a process known as liquefaction because the breakdown of polymers yields a thinner solution. Finally, the dextrins are degraded to glucose and maltose (a glucose dimer) by glucoamylase. The release of simple sugars from a polymer is called saccharification [17].

FERMENTATION OF GLUCOSE TO ETHANOL

The yeast *Saccharomyces cerevisiae* is specialized for fermentation, with approximately 45% of cellular proteins devoted to glycolysis and ethanol fermentation

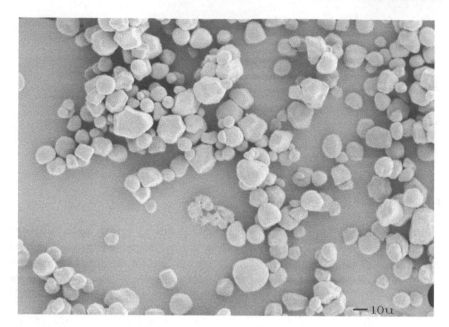

FIGURE 4.2 Starch granules. Granules of standard corn starch are typically 5 to 20 μm in diameter. Scanning electron micrograph courtesy of Victoria L. Finkenstadt.

[18]. Glucose and maltose are fermented to ethanol by *S. cerevisiae* via the same fermentation pathway (Figure 4.3) [19] used to make beverage alcohol. In glycolysis, glucose is converted through a series of reactions to pyruvate, and energy is extracted in the form of four ATP molecules. Then, pyruvate is converted to ethanol in a two-step reaction; pyruvate is decarboxylated to form the more reactive acetaldehyde, which is reduced to ethanol. The second part of the fermentation pathway reoxidizes NADH to NAD$^+$ and thus serves to recover the reducing equivalents that were consumed in the conversion of glucose to pyruvate.

For each glucose fermented, two ethanol and two CO_2 molecules are produced (Table 4.1, Figure 4.3). The theoretical mass yield is only 0.51 g of ethanol per g of fermented glucose. The actual yield is closer to 90–95% of 0.51 g because some glucose is converted to cell mass and side-products such as glycerol, citric acid cycle intermediates, and higher alcohols. Contaminating microorganisms can also lower the yield by converting glucose to other fermentation products such

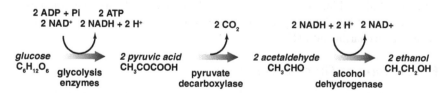

FIGURE 4.3 Fermentation of glucose to ethanol by *Saccharomyces cerevisiae*.

TABLE 4.1
Energy Yield of Fermenting Glucose to Ethanol

	Mass (g)	$\Delta H°_c$ (kJ/mol)[a]
−1 glucose	180	2807
+2 ethanol	2(46)	2(1369)
+2 CO_2	2(44)	0
Sum	0	−69
Yield (ethanol/glucose)	0.51 g/g	0.975 kJ/kJ

[a] Heat of combustion data from Roels [48].

as acetic, lactic, and succinic acids. Because ethanol is used as a fuel, it is also appropriate to consider ethanol yield on an energy basis. The thermodynamic yield can be calculated by comparing the heats of combustion for the products and reactants (Table 4.1). By this measure, converting glucose to ethanol has an amazing theoretical yield of 98–99%, which means that the yeast actually gains little energy benefit from fermenting glucose to ethanol. In other words, ethanol fermentation is an excellent process for generating fuel, because most of the energy from glucose is retained in the fermentation product.

Yeast are ideally suited for use in the fuel ethanol industry. Fermentations run 360 days a year, in tanks containing thousands of gallons of beer, with no pH adjustments and only approximate temperature control (reactors are cooled with well water). As a consequence of the absence of pH control and the production of CO_2, the pH drops steadily during the fermentation and ends up below 4.0. Furthermore, the yeast withstand extreme environmental stresses including high osmolality (beginning solids of 25–30% or higher) and high ethanol concentrations (final concentrations of 12–18% vol.), as well as organic acids produced by contaminating bacteria. The constant contamination of the fermentation is a consequence of the need to run the process in an "open system"— nonaseptically — because the fermentation volumes are quite large and the selling profit margin for ethanol is very low. Fortunately, most bacterial contaminants do not grow below pH 4, and the ability of yeast to do so provides a natural method of suppressing the growth of these contaminants.

Environmental stresses are additive and often synergistic in nature, which means that a combination of many minor stresses, from the perspective of the yeast, equals a single large stress. For example, yeast have reduced tolerance to ethanol at higher temperatures and reduced tolerance to organic acids at lower pH. Despite all of these challenges, *S. cerevisiae* produces ethanol at rates in excess of 3 g $l^{-1} h^{-1}$ and at yields close to 95% of the theoretical maximum. Efforts in the yeast research field are directed at developing strains that produce less glycerol, grow at slightly elevated temperatures (38°C), and withstand even higher ethanol concentrations.

PROCESSES FOR CONVERTING
CORN TO ETHANOL

WET-MILLING AND DRY-GRIND CORN PROCESSES FOR ETHANOL FERMENTATION

Corn is prepared for ethanol fermentation by either wet milling [20] or dry grinding [16] (Figure 4.4). One quarter of the ethanol produced in the United States comes from large-capacity wet-milling plants, which produce ethanol along with a variety of valuable coproducts such as pharmaceuticals, nutriceuticals, organic acids, and solvents. Dry-grind facilities, which account for the remainder of domestic ethanol production, are designed specifically for production of ethanol and animal feed coproducts. Due to the relatively lower capital cost of dry-grind plants and the spread of ethanol plants out of the heart of the U.S. cornbelt, new plants under development and construction are dry-grind facilities.

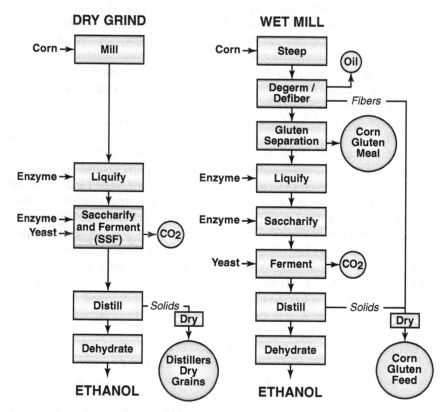

FIGURE 4.4 Wet-milling and dry-grind ethanol production processes. Courtesy of National Corn-to-Ethanol Research Center.

Although both the dry-grind and wet-mill processes produce ethanol, they are very different processes. In dry grinding, dry corn is ground whole and fermented straight through to ethanol. The only coproduct, distillers dry grains with solubles (DDGS), is sold as animal feed. DDGS, which consists of the dried residual materials from the fermentation, contains the nonfermentable parts of the corn and the yeast produced during the fermentation. CO_2 can also be collected and sold to soft-drink producers, but represents a low-profit and limited market.

In wet milling, by contrast, corn kernels are fractionated into each of their major individual components: starch, gluten, germ, and fiber. This imparts two very important advantages compared to dry grinding. First, the parts of the corn can each be marketed separately. So, the germ is used to produce corn oil, the gluten is sold as a high-protein feed to the poultry industry, and the fiber is combined with liquid streams, dried, and sold as a low-protein animal feed. Second, the wet mill produces a pure starch steam, which allows for the starch to be made into numerous different products. In addition to being fermented to ethanol, the starch can be modified for use in textiles, paper, adhesives, or food. Maltodextrins and high-fructose corn syrup, the major sweetener used by the U.S. food industry, are made enzymatically from starch. The starch can also be converted enzymatically to a fairly pure glucose stream and then fermented to any number of products. A partial list includes amino acids, vitamins, artificial sweeteners, citric acid, and lactic acid, in addition to ethanol. If ethanol is produced, the yeast can be spray-dried and marketed as distillers yeast, a high-protein, low-fiber product suitable for feeding animals and fishes. Although no wet mill makes all of these products, it is not unusual for large facilities to have multiple starch product streams.

Dry-grind plants do not have the capability to ferment corn starch to these products in part because the additional products are nonvolatile and, therefore, cannot be simply separated by distillation from all of the other material in the fermentation. In summary, a wet mill that converts all its starch to ethanol produces at least two or three additional high-value products compared to a dry-grind facility. Of course, these additional products are realized only with much higher capital expenses. As discussed later in this chapter, there are several efforts underway to develop less capital-intensive processes for either totally or partial fractionating corn that would be suitable for implementation at dry-grind facilities.

DRY-GRIND ETHANOL PRODUCTION

Starch Conversion

Corn is received at the plant and separated from the chaff, and the kernels are milled to a coarse flour. Particle sizing is a compromise between grinding fine enough to provide increased surface area (to make starch granules available for swelling and hydrolysis), yet large enough to allow separation of residual solids from the liquid. The corn meal is mixed with water, and the resulting mash is adjusted to pH 6 and then mixed with alpha-amylase. The mash is heated above

110°C in a jet cooker using direct steam. A jet cooker is in essence a pipe with a narrowing and a steam inlet directly upstream. The narrowing is carefully engineered to provide maximum mixing of the starch slurry with steam, and also to cause shearing, which aids in thinning the starch. Upon exiting the jet cooker, the corn slurry enters a holding column where the mixture is kept at 110°C for 15 minutes. From the holding column, the slurry enters a flash tank at atmospheric pressure and 80–90°C. Additional alpha-amylase is added and the mash is liquefied for approximately 30 minutes. The jet cooking and liquefaction steps break apart the starch granules and reduce the size of the polymers. The shorter molecules, termed dextrins, contain approximately five to ten glucose molecules [21]. Subsequently, the liquefied mash is cooled to 32°C and the pH is lowered to 4.5–5.0 using phosphoric acid and recycled backset from the bottom of the ethanol distillation column.

Fermentation

Dry yeast is hydrated or conditioned and then added to the mash along with glucoamylase, to initiate simultaneous saccharification and fermentation (SSF). Glucoamylase cleaves the dextrins at α-1,4-glucosidic linkages, releasing glucose and maltose for yeast fermentation. The SSF process reduces the extent of microbial contamination because glucose is consumed by yeast as it is formed. The SSF process also reduces osmotic stress, because the yeast cells are exposed to a relatively lower sugar concentration. The dry-grind ethanol fermentation process lasts for 48 to 72 h and yields approximately 2.7 gallons of ethanol per bushel of corn.

Distillation and Dehydration

At the end of fermentation, the beer contains 10% or more ethanol by volume. Separation of ethanol from the whole fermentation mixture begins with distillation on a beer column [22]. Further removal of water is accomplished in subsequent distillation steps using a rectifier and/or stripper. Conventional distillation/rectification methods yield 95% pure ethanol, at which concentration of ethanol and water form an azeotrope. The remaining 5% water is removed by molecular sieves, which rely on size exclusion to separate the smaller ethanol molecules from water [23]. Finally, anhydrous (100% or 200 proof) ethanol is denatured, typically with 5% gasoline or with higher-chain alcohols formed in the fermentation, to exempt the ethanol from beverage alcohol taxes.

Stillage Processing and Feed Products

The slurry remaining after distillation of ethanol, known as stillage, is concentrated by centrifugation. The solids cake is referred to as corn distillers grains. Up to one third of the liquid fraction, known as thin stillage, is recycled (backset) into the mash. The remaining liquid is concentrated by evaporation and mixed with the corn distillers grains. The mixture is either sold as wet distillers grains or dried to generate DDGS. The moisture content and correspondingly short shelf-life of wet distillers grains limit use of this feed product to the immediate vicinity of ethanol plants, though the shelf-life can be lengthened by adding organic acids as preservatives.

ETHANOL PRODUCTION BY WET MILLING

In contrast to the dry-grind process (where the whole corn kernel is ground and enters the fermenter) in wet milling, the kernel is first fractionated into separate components and only the starch enters the fermenter. Corn wet milling plants are often referred to as biorefineries, comparable to petroleum refineries, because wet milling fractionates corn into its components and then processes the components into more valuable products [20]. Wet milling separates the kernel into germ (oil), gluten (protein), fiber, and starch fractions, yielding corn oil, animal feed, and a variety of products derived from starch.

Steeping

The wet-milling process starts with steeping, which enables isolation of the different kernel fractions [20]. Corn is screened to remove foreign material, and soaked in dilute (0.12–0.20%) sulfurous acid at 52°C for (typically) 30–36 h. *Lactobacillus* and related bacteria growing in the steep water produce lactic acid and other metabolites that further acidify the medium. Steeping occurs in a series of large stainless steel tanks, with steep water recirculated countercurrently from tanks holding "older" corn that is nearing the end of the steeping process to "newer" corn that is beginning the steep. The effect of recycling the steep liquid is the progression of the corn up a SO_2 concentration gradient. The combined action of sulfurous acid and lactic acid, and probably also direct effects of microbes growing in the steep, prepare the kernels for processing into fractions. Steeping softens and swells the kernels, disrupts disulfide bonds between the protein and starch in the endosperm, and releases the starch granules into solution.

Oil, Fiber, and Gluten Separation

After steeping, the germ (which contains most of the corn oil) is dislodged from the kernel by gentle disruption using a germ mill. The germ fraction is separated in hydroclones based upon its low density (high oil content) and then washed to remove loose starch and gluten. The germ is pressed and dried, and the oil is either extracted on-site or sold to a corn oil refiner. If the oil is extracted on-site, the residual material is blended into corn gluten feed. The slurry exiting the hydroclones is screened to separate fiber from protein (gluten) and starch. Fiber is repeatedly washed and the destarched fiber is incorporated with steepwater solids and the bottoms of the distillation column into corn gluten feed. Gluten is separated from starch by centrifugation and dried to produce corn gluten meal. Corn steep liquor, the liquid remaining after removal of kernel components, can also be sold separately as either a feed or fermentation ingredient.

Starch Conversion

Starch is washed to remove residual protein and is converted to a glucose syrup. First, the starch is jet-cooked and held for liquefaction at 90°C with alpha-amylase. In contrast to the dry-grind process, all of the alpha-amylase is added prior to jet cooking. Then glucoamylase and pullulanase, a α-1→6 debranching enzyme, are added to convert dextrin polymers to sugars. Addition of pullulanase ensures good conversion of dextrins to glucose by decreasing formation of isomaltose, a glucose dimer. Isomaltose is present at starch polymer branching points and can also be formed by "reversion" of glucose to isomaltose in a reaction catalyzed by glucoamylase. After saccharification, glucose is fermented to ethanol by an industrial strain of *S. cerevisiae*. In wet mills, the fermentation is often run using a series of fermenters in a semi-continuous process. Approximately 2.5 gallons of ethanol are produced from a bushel of corn in the wet-milling process.

Coproducts from Wet Milling

As stated previously, wet-milling operations have the capacity and flexibility to make more products than a dry-grind ethanol facility, because the individual parts of the kernel are fractionated. Products resulting directly from wet milling are corn oil for cooking, CO_2, which may be captured and sold for carbonation of beverages, and corn gluten meal and corn gluten feed, which are sold as animal feeds. Additional products may be obtained from starch, by siphoning part of the sugar stream into alternate products. The product mix from a wet mill can be changed (within limits) in response to market conditions, and has grown to include products that were formerly synthesized by chemical processes. Alternative fermentation products include organic acids, amino acids, sugar alcohols, polysaccharides, pharmaceuticals, nutraceuticals, fibers, biodegradable films, solvents, pigments, enzymes, polyols, and vitamins [24]. The simple sugars derived from starch can also be converted enzymatically to sweeteners including high fructose corn syrup, which is the primary food use of corn in the United States.

FUTURE DIRECTIONS IN THE CORN
ETHANOL INDUSTRY

There are a number of trends shaping the industry's future. Wet millers will continue to develop value-added products for the starch stream and alternative uses for corn fiber. The dry-grind industry is looking at fractionating the corn prior to fermentation to realize better value from the corn, and at basic process modifications for lowering energy use [25, 26]. As is the case for corn fiber derived from wet milling, the dry-grind industry is also seeking alternate uses for its lowest value product, DDGS.

ALTERNATIVE FEEDSTOCKS

Corn Fiber

Wet mills have been investigating converting their least valuable residue from corn processing (hulls and deoiled germ, known as corn fibers) into ethanol. Corn fiber is an attractive target for value-added research, because its value could be increased by converting it to ethanol instead of adding it to animal feed. The fiber is generated as a by-product at wet-milling facilities, and so there is no added cost for collection and transport of the material. Corn fiber contains 11–23% residual starch from wet milling and 12–18% cellulose (w/w, dry basis); the glucose in both polymers could be fermented to ethanol by traditional yeast. In addition, corn fiber contains 18–28% xylan and 11–19% arabinan. However, industrial yeast strains currently used for fermenting corn starch do not ferment arabinose and xylose, and the few naturally occurring yeast that do ferment pentoses produce low ethanol yields. Consequently, genetically engineered microorganisms will be required for efficient conversion of pentose sugars to ethanol [27, 28]. A method to ferment the fiber to ethanol would increase ethanol yield from corn by 10%, while also generating feed coproducts with higher protein content [29]. Fermentation of corn fiber is being evaluated by Aventine Renewable Energy (Pekin, IL) to establish the process economics and robustness [30].

Corn Stover and Alternative Starches

When corn is harvested in the field, only the grain is collected for transport and sale. However, the rest of the corn plant — stalks, cobs, etc. referred to as stover — also contains carbohydrates (58% wt/wt) that could potentially serve as a fermentation substrate [31]. For each pound of harvested corn grain, 1.0-1.5 pounds of stover are produced, and some estimate that about half of this could be harvested without negatively affecting soil quality. However, a number of constraints limit utilization of this type of feedstock, and currently there are no commercial plants that convert corn fiber or stover to ethanol. Instead, growth in production of ethanol outside the U.S. corn belt may be driven in the near term by use of alternate starch feedstocks. The Renewable Fuels Association reports that 12% of the U.S. domestic sorghum crop was fermented to ethanol in 2004. Starch from wheat, barley, rye, and cassava, and sucrose from sugar cane and sugar beets, are also fermented to ethanol around the world. Other crops such as hulless barley, field peas, and even cattails have been explored as fermentation feedstocks. In addition, ethanol fermentation can be used as an alternative to waste disposal for residues such as whey and potato processing solids in some localities.

Other Types of Biomass

Displacement of petroleum by fuel ethanol is approaching 3% of the liquid transportation fuel used in the United States. Expanding ethanol to replace more than 10% of fuel needs will require development of additional and lower-cost feedstocks. Only lignocellulosic biomass is available in sufficient quantities to augment starch as an ethanol feedstock source. As discussed previously, corn fiber and corn stover are potential sources of lignocellulosic biomass for fermentation. Other possible feedstocks are agricultural residues such as wheat and rice straws and sugar cane bagasse, energy crops including switch grass and softwood trees, and waste materials such as pulp and paper sludge and recycled office paper. The capacity to process and ferment even one of these categories of biomass would significantly increase production of ethanol; however, practical aspects of collection and storage must be addressed for many of these resources.

Several technological constraints limit fermentation of biomass feedstocks. Lignocellulosic biomass can be pretreated and enzymatically hydrolyzed to yield a mixture of sugars including glucose, galactose, arabinose, and xylose [32]. However, hydrolytic enzymes are inefficient and expensive. More-effective pretreatment methods, as well as active and cost-effective enzymes, are needed for an economical process. As mentioned previously, microbes that efficiently ferment multiple sugars to ethanol must be developed in order to convert biomass to ethanol. Fermenting microbes also must tolerate the inhibitory compounds generated during biomass hydrolysis, or alternatively, cost-effective methods for inhibitor abatement must be in place. A study comparing dry-grind production of ethanol from corn and ethanol produced from corn stalks concluded that producing ethanol from corn stover would cost $1.45 per gallon compared to $0.96 from corn starch [33]. Despite these obstacles, one company, Iogen Corp. (Ottawa) has begun to produce ethanol from biomass.

NEW PRODUCTS FROM WET-MILLED CORN STARCH

Corn wet mills have a long history of converting starch to a wide variety of products in addition to or instead of ethanol. As described above in the discussion of wet milling, starch products from wet milling also have many applications beyond the food and beverage industries in pharmaceutical, cosmetic, paper, and packaging industries. Recently, corn wet millers have begun to adopt processes for converting starch into biodegradable polymers. Such products represent a fundamental shift from other nonethanol products, because they directly compete with petroleum-based products and have the potential for virtually unlimited growth. A number of corn processors in the United States have a biopolymer either on the market or under development. Cargill Dow produces PLA (polylactide) polymer fiber under the trade name NatureWorks, for use in packaging, films, and resins. ADM and Metabolix have formed an alliance to scale-up and commercialize PHA (polyhydroxyalkanoate) polymers, for marketing as renewable alternatives to traditional petrochemical plastics used in making molded goods, films, and coated papers. A joint venture between DuPont and Tate and Lyle was formed to produce 1,3-propanediol (Bio-PDO™) from corn starch as an alternative to petrochemically-derived PDO. Sorona®, a family of polymers made from PDO, is used in fibers and fabrics, films, and resins.

MODIFYING THE CORN DRY-GRIND PROCESS

Quick Fiber and Quick Germ Processes

As mentioned earlier in the discussion of dry-grind technology, the dry-grind process is being adapted to capture some of the advantages of the wet-milling system. The Quick Germ [34] and Quick Fiber [35] fractionation methods increase processing efficiency as well as the value of feed coproducts. These methods add technology to the beginning of the dry-grind process, removing the germ and fiber fractions of the corn kernel prior to starch processing. First, the corn is soaked in water for 3–12 hours to hydrate the germ, which is recovered by density separation. Next, the specific gravity of the mash is adjusted and pericarp fiber is recovered using hydrocyclones. The starch is then fermented in the traditional dry-grind process, and the germ and fiber can be processed separately into other value-added products: corn oil (from germ) and corn fiber oil and corn fiber gum (from corn fiber). Corn fiber oil, which is distinct from traditional corn germ oil, is potentially a valuable coproduct because it contains phytosterols known to have cholesterol-lowering properties. Another process, known as enzymatic milling or the E-mill process, refers to the addition of proteases and amylases prior to germ separation [36]. E-milling allows recovery of the gluten, while avoiding a full steeping process and the health and environmental concerns associated with sulfite.

Because the germ and fiber do not enter the fermenter, the fermentation residuals have lower fiber content and correspondingly higher protein content, resulting in a higher-value feed product. On the process side, these modified milling technologies increase the effective capacity of fermentation tanks, because removal of the germ and fiber frees reactor volume for additional starch. Bringing these modified processes to the dry-grind plant would therefore increase the facility's ethanol production by 8 to 27% [37].

Very-High-Gravity Fermentations

A laboratory process to ferment highly concentrated mash, with greater than 30% solids, has been developed [38]. Very-high-gravity (VHG) fermentations produce 21 to 23% ethanol under optimal fermentation conditions. The VHG process requires a mash with high sugar concentration and low viscosity, which can be achieved by adding enzymes (e.g., proteases, glucanases, and amylases) or by double mashing, in which the solids are removed from an initial mash, and the liquids are used to prepare the second (VHG) extract. VHG fermentations use less water than conventional fermentations. Energy costs also decrease because there is less volume to cool for fermentation and then distill.

New Processing Technology to Decrease Energy Use

"Cold saccharification" technology allows enzymatic release of glucose from starch without liquefaction (e.g., jet-cooking) with steam. Laboratory-scale alcohol fermentation of ground rice without cooking was reported in 1963 [39] and an industrial-scale fermentation was reported in 1982 [40]. Use of very high solid concentrations improved productivity and yielded high concentrations of ethanol. In 2004, commercial technologies for direct conversion of raw starch to ethanol were developed for use in modern dry-grind ethanol fermentation facilities [41, 42]. A key to cold-process technology is the development of robust and efficient conversion enzymes [43]. The benefits of a no-cook process include reduced energy, water, and waste costs, reduced capital and related maintenance expenses, improved conversion efficiency resulting in increased ethanol yield, and increased protein content and quality of feed coproducts. Possible drawbacks to the technology include the cost and amount of enzyme required for the process and an increased chance for microbial contamination and corresponding loss of yield, because heating to partially pasteurize the mash does not occur. Heating that occurs in the standard dry-grind ethanol process also aids release of starch that is bound to fiber or protein, and inactivates some toxins that may be present in the grain. The no-cook method must manage these issues by means of alternate technologies [44]. Widespread adoption of cold-hydrolysis technology stands to greatly impact the productivity and profitability of the ethanol industry.

ALTERNATE USES FOR DDGS

With regard to feed coproducts, there is a need to diversify the markets for DDGS. Research aimed at modifying the amino acid composition, protein composition, and phosphorous content of DDGS should result in higher quality, consistent composition of feed ingredients and encourage expanded use of DDGS in poultry and swine rations. Nonfeed uses for DDGS have also been developed, prompted by the increased availability of DDGS resulting from increased ethanol production. Deicers, cat litter, and lightweight "ag-fiber" shipping containers can be produced from DDGS. DDGS could also be used to produce biogas, which could be recovered and used on-site to fuel the plant [26]. A plant fueled in this way would be less dependent on feed selling prices and power and natural gas purchase prices.

ZEIN PROTEIN FROM CORN DRY GRINDING OR WET MILLING

Zein is a biodegradable resin that has value for food and cosmetic uses, however, currently available methods for recovering and purifying zein protein from corn are too expensive to compete with petroleum-based films and plastics. The COPE (corn oil and protein extraction) process is being developed to inexpensively extract zein protein from either milled corn or DDGS, for use in the plastic and film-packaging industries [45]. Achieving cost-effective methods for extraction of zein and synthesis of zein-based biodegradable polymers could add value to DDGS and the nonstarch fraction of corn.

HYBRID AND STRAIN DEVELOPMENT

"Self-processing" hybrid corn kernels are being developed. This type of hybrid accumulates starch-hydrolyzing enzymes in the endosperm of transgenic corn kernels [46]. Work has also been carried out to construct fermenting yeast strains with built-in amylolytic activity [47]. Using these strategies, genes for enzymes with known properties and specificities could be used to engineer customized grains and/or yeast. Self-processing grains or starch-degrading and fermenting yeast would thus have built-in enzymatic activities designed to meet specific processing requirements. Other genetically modified corn hybrids are being developed to generate alternate starches and complex carbohydrates. Starch from these hybrids would have improved gelling properties, viscosity, flavor, stability, adhesion, film formation, or properties that enhance the efficiency of starch processing.

CONCLUSIONS

Use of ethanol, a renewable transportation fuel, is expected to expand significantly because of concerns regarding the environment and energy security. The U.S. ethanol industry will continue to utilize corn as its primary source of fermentable sugar. New feedstocks, technologies, and products moving from the laboratory to the marketplace will increase the productivity and viability of the fuel ethanol industry.

REFERENCES

1. Berg, C., *World Fuel Ethanol Anaylsis and Outlook*, F.O. Licht, Kent, U.K., 2004.
2. Kovarik, B., Henry Ford, Charles F. Kettering and the fuel of the future, *Automot. Hist. Rev.*, 32, 7–27, 1998.
3. Shapouri, H., Duffield, J.A. and Graboski, M.S., Estimating the net energy balance of corn ethanol, *Agric. Econ. Rep.* 721. U.S. Department of Agriculture, Washington, DC, 1995.
4. Kim, S. and Dale, B.E., Allocation procedure in ethanol production system from corn grain. I. System expansion, *Int. J. Life Cycle Assessment*, 7, 237–243, 2002.
5. Wang, M., Saricks, C. and Santini, D., Effects of fuel ethanol use on fuel-cycle energy and greenhouse gas emissions. ANL/ESD-38, U.S. Department of Energy Argonne National Laboratory Center for Transportion Research, 1999.
6. Shapouri, H. and McAloon, A., The 2001 net energy balance of corn-ethanol, In *Corn Utilization and Technology Conference Proceedings*, Tumbleson, M., Ed., Nat. Corn Growers Assoc., Indianapolis, 2004.
7. Shapouri, H., Duffield, J. A. and Wang, M. The energy balance of corn ethanol: an update, *Agric. Econ. Rep.* 813, 2002.
8. Urbanchuk, J.M. and Kapell, J., *Ethanol and the Local Community*, AUS Consultants, SJH & Company, Moorestown, NJ, Boston, 2002.
9. Kotrba, R., State legislation roundup and review, *Ethanol Prod. Mag.*, 11, 34–41, 52–55, 2005.

10. Watson, S. A., Description, development, structure, and composition of the corn kernel, In *Corn: Chemistry and Technology*, 2nd Ed., White, P.J. and Johnson, L.A., Eds. American Association of Cereal Chemists, St. Paul, MN, 2003, 69–106.

11. Haefele, D.; Owens, F., O'Bryan, K. and Sevenich, D., Optimization of corn hybrids for fuel ethanol production., In *Proceedings of the ASTA 59th Annual Corn and Sorghum Research Conference, CD-ROM*, Amer. Seed Trade Assoc., Alexandria, VA, 2004.

12. Dien, B.S., Bothast, R.J., Iten, L.B., Barrios, L. and Eckhoff, S.R., Fate of Bt protein and influence of corn hybrid on ethanol production., *Cereal Chem.*, 79, 582–585, 2002.

13. Bothast, R.J., New technologies in biofuel production., *Agric. Outlook Forum*, 2005.

14. Anonymous, New Pioneer, NCGA technology available, *Ethanol Prod. Mag.*, 10, 10, 2004.

15. Boyer, C.D. and Shannon, J.C., Carbohydrates of the kernel., In *Corn Chemistry and Technology*, 2nd Ed., White, P.J. and Johnson, L.A., Eds., American Association of Cereal Chemists, St. Paul, MN, 2003, pp. 289–311.

16. Kelsall, D.R. and Lyons, T.P., Grain dry milling and cooking procedures: Extracting sugars in preparation for fermentation, In *The Alcohol Textbook*, 4th Ed., Jacques, K.A., Lyons, T.P. and Kelsall, D.R., Eds., Nottingham University Press, Nottingham, U.K., 2003, pp. 9–21.

17. Power, R.F., Enzymatic conversion of starch to fermentable sugars., In *The Alcohol Textbook*, 4th Ed., Jacques, K.A., Lyons, T.P. and Kelsall, D.R., Eds. Nottingham University Press, Nottingham, U.K., 2003, pp. 23–32.

18. Futcher, B., Latter, G.I., Monardo, P., McLaughlin, C.S. and Garrels, J.I., A sampling of the yeast proteome, *Molec. Cell. Biol.* 19, 7357–7368, 1999.

19. Russell, I., Understanding yeast fundamentals, In *The Alcohol Textboo*, 4th Ed., Jacques, K.A., Lyons, T.P., and Kelsall, D.R., Eds., Nottingham University Press, Nottingham, U.K., 2003, pp. 85–119.

20. Johnson, L.A. and May, J.B., Wet milling: The basis for corn biorefineries, In *Corn: Chemistry and Technology*, 2nd Ed., White, P.J. and Johnson, L.A., Eds., American Association of Cereal Chemists, St. Paul, MN, 2003, pp. 449–494.

21. Kennedy, J.F., Knill, C.J. and Taylor, D.W., Maltodextrins, In *Handbook of Starch Hydrolysis Products and Their Derivatives*, Kearsley, M.W. and Dziedzic, S.Z., Eds., Blackie Academic & Professional: Glasgow, 1995, pp. 65–82.

22. Madson, P.W., Ethanol distillation: the fundamentals, In *The Alcohol Textbook*, 4th Ed., Jacques, K.A., Lyons, T.P. and Kelsall, D.R., Eds., University of Nottingham Press, Nottingham, U.K., 2003, pp. 319–336.

23. Bibb Swain, R.L., Development and operation of the molecular sieve: an industry standard, In *The Alcohol Textbook*, 4th Ed., Jacques, K.A., Lyons, T.P. and Kelsall, D.R., Eds., University of Nottingham Press: Nottingham, U.K., 2003, pp. 337–341.

24. Leathers, T.D., Upgrading fuel ethanol coproducts, *SIM News* 48, 210–217, 1998.

25. Dien, B.S., Bothast, R.J., Nichols, N.N. and Cotta, M.A., The U.S. corn ethanol industry: An overview of current technology and future prospects, *Int. Sugar J.*, 104, 204–211, 2002.

26. Belcher, A., The world looks to higher-tech to advance fuel ethanol production into the 21st century, *Int. Sugar J.*, 107, 196–199, 2005.

27. Dien, B.S., Cotta, M.A. and Jeffries, T.W., Bacteria engineered for fuel ethanol production: Current status, *Appl. Microbiol. Biotechnol.*, 63, 258–266, 2003.

28. Jeffries, T.W. and Jin, Y.-S., Metabolic engineering for improved fermentation of pentoses by yeast, *Appl. Microbiol. Biotechnol.*, 63, 495–509, 2004.
29. Gulati, M., Kohlmann, K., Ladisch, M.R., Hespell, R.B. and Bothast, R.J., Assessment of ethanol production options for corn products, *Bioresource Tech.*, 58, 253–264, 1997.
30. Mosier, N.S., Hendrickson, R., Brewer, M., Ho, N., Sedlack, M., Dreshel, R., Welch, G., Dien, B.S., Aden, A. and Ladisch, M.R., Industrial scale-up of pH-controlled liquid hot water pretreatment of corn fiber for fuel ethanol production, *Appl. Biochem. Biotechnol.*, 125, 77–98, 2005.
31. Wyman, C.E., Applications of corn stover and fiber, In *Corn Chemistry and Technology*, 2nd Ed., White, P.J. and Johnson, L.A., Eds., American Association of Cereal Chemists, St. Paul, MN, 2003, pp. 723–750.
32. Grohmann, K. and Bothast, R.J., Saccharification of corn fibre by combined treatment with dilute suphuric acid and enzymes, *Process Biochem.*, 32, 405–415, 1997.
33. Wallace, R., Ibsen, K., McAloon, A. and Yee, W., *Feasibility study for co-locating and integrating ethanol production plants from starch and lignocellulosic feedstock*, U.S. Department of Agriculture and U.S. Department of Energy, National Renewable Energy Laboratory Technical Report, 2004.
34. Singh, V. and Eckhoff, S.R., Effect of soak time, soak temperature, and lactic acid on germ recovery parameters, *Cereal Chem.*, 73, 716–720, 1996.
35. Singh, V., Moreau, R.A., Doner, L. W., Eckhoff, S.R. and Hicks, K.B., Recovery of fiber in the corn dry grind ethanol process: a feedstock for valuable coproducts, *Cereal Chem.*, 76, 868–872, 1999.
36. Johnston, D.B. and Singh, V., Enzymatic milling of corn: Optimization of soaking, grinding, and enzyme incubation steps, *Cereal Chem.*, 81, 626–632, 2004.
37. Singh, V., Johnston, D.B., Naidu, K., Rausch, K.D., Belyea, R.L. and Tumbleson, M.E., Comparison of modified dry-grind corn processes for fermentation characteristics and DDGS composition, *Cereal Chem.*, 82, 187–190, 2005.
38. Thomas, K.C., Hynes, S.H. and Ingledew, W.M., Practical and theoretical considerations in the production of high concentrations of alcohol by fermentation, *Process Biochem.*, 31, 321–331, 1996.
39. Yamasaki, I., Ueda, S. and Shimada, T., Alcoholic fermentation of rice without previous cooking by using black-koji amylase, *J. Ferm. Assoc. Jpn.*, 21, 83–86, 1963.
40. Matsumoto, N., Fukushi, O., Miyanaga, M., Kakihara, K., Nakajima, E. and Yoshizumi, H. Industrialization of a noncooking system for alcoholic fermenation from grains, *Agric. Biol. Chem.*, 46, 1549–1558, 1982.
41. Swanson, T., Partnering in progress, *Ethanol Prod. Mag.*, 10, 62–68, 2004.
42. Berven, D., The making of Broin Project X, *Ethanol Prod. Mag.*, 11, 67–71, 2005.
43. Wong, D. and Robertson, G., Applying combinatorial chemistry and biology to food research, *J. Agric. Food Chem.*, 52, 7187–7198, 2004.
44. Galvez, A., Analyzing cold enzyme starch hydrolysis technology in new ethanol plant design, *Ethanol Prod. Mag.*, 11, 58–60, 2005.
45. Cheryan, M., Corn oil and protein extraction method, U.S. Patent 6,433,146, 2002.
46. Craig, J.A., Batie, J., Chen, W., Freeland, S.B., Kinkima, M. and Lanahan, M.B., Expression of starch hydrolyzing enzymes in corn, In *Corn Utilization and Technology Conference Proceedings*, Tumbleson, M., Ed., National Corn Growers Association and Corn Refiners Association, Indianapolis, 2004.

47. Shigechi, H., Koh, J., Fujita, Y., Matsumoto, T., Bito, Y., Ueda, M., Satoh, E., Fukuda, H. and Kondo, A., Direct production of ethanol from raw corn starch via fermentation by use of a novel surface-engineered yeast strain codisplaying glucoamylase and α-amylase, *Appl. Environ. Microbiol.*, 70, 5037–5040, 2004.
48. Roels, J.A., Macroscopic theory, and microbial growth and product formation, In *Energetics and Kinetics in Biotechnology*, Elsevier Biomedical Press, Amsterdam, 1983, pp. 23–73.

5 Development of Alfalfa (*Medicago sativa* L.) as a Feedstock for Production of Ethanol and Other Bioproducts

Deborah A. Samac, Hans-Joachim G. Jung, and JoAnn F. S. Lamb
USDA-ARS-Plant Science Research, University of Minnesota, St. Paul

CONTENTS

Alfalfa (*Medicago sativa* L.) has considerable potential as a feedstock for production of fuels, feed, and industrial materials. However, unlike other major field crops such as corn and soybeans, which are commonly refined for production of fuel and industrial materials, refining of alfalfa remains undeveloped. Instead, alfalfa is primarily processed and used on-farm in the form of dried hay, silage, and fresh forage known as "greenchop," or is grazed by animals in pastures. In many countries, including the United States, alfalfa is used as a basic component in feeding programs for dairy cattle and is an important feed for beef cattle, horses, sheep, and other livestock. Known as the "Queen of the Forages," alfalfa provides highly nutritious forage in terms of protein, fiber, vitamins, and minerals for ruminant animals. If alfalfa is developed to its full potential as a feedstock for biorefining, a major shift may occur in the manner in which alfalfa is produced and used for feeding farm animals.

CURRENT ALFALFA CULTIVATION AND UTILIZATION

A number of attributes make alfalfa an attractive crop for production of biofuels and for biorefining. Alfalfa has a long history of cultivation around the world. It was introduced several times into North America during the 1700s and 1800s and is currently grown across the continent (Russelle, 2001). In the United States, alfalfa is the fourth most widely grown crop with over 9.3 million hectares of alfalfa harvested in 2003 (USDA-NASS, 2004). It is a perennial plant that is typically harvested for four years (an establishment year plus three subsequent years). Depending on location, alfalfa is harvested three or more times each year by cutting the stems near ground level. On average across the United States, alfalfa yields 7.8 Mg of dry matter (DM) per hectare each year, although yields can vary by location from 3.4 (North Dakota) to 18.4 (Arizona) Mg ha^{-1} (USDA-NASS, 2004). In 2003 the national harvest of alfalfa was over 69 million metric tons (USDA-NASS, 2004). The technology for cultivation, harvesting, and storing alfalfa is well established, machinery for harvesting alfalfa is widely available, and farmers are familiar with alfalfa production. There is a well-developed industry for alfalfa cultivar development, seed production, processing, and distribution. Alfalfa breeders have utilized the extensive germplasm resources of alfalfa to introduce disease and insect resistance, expand environmental adaptation, and improve forage quality. Nonetheless, alfalfa cultivation requires fertile, deep, well-drained soils of near neutral pH and is limited to humid areas with adequate rainfall. In arid or semi-arid areas, irrigation is essential for crop production. Despite breeding efforts that have increased disease and pest resistance, alfalfa yields have not increased substantially over the past 25 years (Brummer, 1999).

The high biomass potential of alfalfa is based on underground, typically unobserved traits. Alfalfa develops an extensive, well-branched root system that is capable of penetrating deep into the soil. Root growth rates of 1.8 m a year are typical in loose soils (Johnson et al., 1996) and metabolically active alfalfa roots have been found 18 m or more below ground level (Kiesselbach et al., 1929). This deep root system allows alfalfa plants to access water and nutrients that are not available to more shallowly rooted annual plants, which enables established alfalfa plants to produce adequate yields under less than optimal rainfall conditions. Alfalfa roots engage in a symbiotic relationship with the soil bacterium *Sinorhizobium meliloti*. This partnership between the plant and bacterium results in the formation of a unique organ, the root nodule, in which the bacterium is localized. The bacteria in root nodules take up nitrogen gas (N_2) and "fix" it into ammonia. The ammonia is assimilated through the action of plant enzymes to form glutamine and glutamate. The nitrogen-containing amide group is subsequently transferred to aspartate and asparagine for transport throughout the plant. On average, alfalfa fixes approximately 152 kg N_2 ha^{-1} on an annual basis as a result of biological nitrogen fixation, which eliminates the need for applied nitrogen fertilizers (Russelle and Birr, 2004). Although a significant proportion of the fixed nitrogen is removed by forage harvest, fixed nitrogen is also returned to the soil for use by subsequent crops. This attribute of increasing soil fertility has made alfalfa and other plants in the legume family crucial components of agricultural systems worldwide. Cultivation of alfalfa has also been shown to improve soil quality, increase organic matter, and promote water penetration into soil.

Responsible stewardship of agricultural lands has never been more important. Utilization of alfalfa as a biomass crop has numerous environmental advantages. There is an urgent need to increase the use of perennials in agricultural systems to decrease erosion and water contamination. Annual row crop production has been shown to be a major source of sediment, nutrient (nitrogen and phosphorus), and pesticide contamination of surface and ground water. Perennial crops such as alfalfa can reduce the nitrate concentrations in soil and drainage water, and prevent soil erosion (Huggins et al., 2001). In addition, energy costs associated with production of alfalfa are low. A recent study shows that energy inputs for production of alfalfa are far lower than for production of corn and soybean, and very similar to switchgrass (Kim and Dale, 2004), primarily because alfalfa does not require nitrogen fertilizer. Biorefining could increase the return on alfalfa production so that cultivation of the crop is more economically attractive, as well as environmentally beneficial.

An additional advantage of using alfalfa for biofuel production compared to other crops is the ability to easily separate leaves and stems to produce co-products. In fact, alfalfa herbage can almost be considered two separate crops because leaves and stems differ so dramatically in composition. On a dry weight basis, total alfalfa herbage contains 18–22% protein with leaves containing 26–30% protein and stems only 10–12% (Arinze et al., 2003). In some analyses, alfalfa protein has been valued highly, theoretically greatly reducing the cost of the lignocelluose fraction (Dale, 1983). Several different integrated processes for refining alfalfa have been proposed based primarily on the method of refining the protein fraction. From field-dried hay, leaves may be separated from stem material mechanically (see "Protein and Fiber Separation" below). The leaf meal could be used as a high-protein feed with the stems utilized for gasification and conversion to electricity (Downing et al., 2005) or fermentation to ethanol (Dale, 1983). Alternatively, protein could be extracted from total ground material and the residue used for fermentation. Fresh forage can be "juiced" to remove protein and the residue fermented to ethanol or other products (Koegel et al., 1999; Sreenath et al., 2001; Weimer et al., 2005). An economic analysis of these alternatives is beyond the scope of this chapter. However, a comparison of the potential costs and revenues of different biobased feedstocks to produce ethanol and other products is clearly needed to advance biomass refining from the theoretical to practical stages.

DEVELOPMENT AND CULTIVATION OF ALFALFA FOR BIOMASS

Genetic modification to improve alfalfa over the past century has increased resistance to several diseases and pests and widened the range of environmental adaptation of the crop by producing varieties that differ in fall dormancy and winter hardiness. Most improvements in forage quality of alfalfa have occurred through changes in harvest management and production practices. Alfalfa produced as feed for ruminant livestock is harvested frequently at early maturity when the leaf to stem ratio is high, producing hay that is high in protein and easily digested. Maximum forage yield, which occurs at later maturity stages in alfalfa, is usually sacrificed in order to produce high-quality hay. For competitive use of alfalfa as a biofuel feedstock, research is needed to develop alfalfa germplasm and management strategies that yield more biomass (both leaf and stem) with minimal production costs.

Marquez-Ortiz et al. (1999) reported that individual stem diameter was heritable and controlled by additive genetic effects and suggested that selection for larger stems in alfalfa was feasible. Volenec et al. (1987) found that selection for high yield per stem may be an effective means to increase forage yield, but plants may have less digestible, larger stems. Germplasms from southern Europe referred to as Flemish types are a genetic source for large stem size and resistance to foliar diseases in alfalfa, but display early maturity, lack winter hardiness, and are susceptible to root and crown diseases (Barnes et al., 1977).

The effects of plant population or density on stem, leaf and total forage yield have been well documented in alfalfa. As alfalfa plant densities increase, annual forage yield per land area unit increases, but yield of individual alfalfa stems and number of stems per plant decreases (Cowett and Sprague, 1962; Rumbaugh 1963). Hansen and Krueger (1973) reported that higher plant densities produced finer stems, decreased root and crown weights and increased leaf drop due to shading. Volenec et al. (1987) stated that stem diameter and nodes per stem decreased as plant density increased and that shoot weight was an important component of plant weight, especially at high plant densities. Decreasing plant density to approximately 45% (180 plants m^{-2}) of that conventionally used in alfalfa hay production stands (450 plants m^{-2}) and delaying harvest until the green pod stage maximized leaf and stem yield in four unrelated alfalfa germplasms (Figure 5.1). The reduced plant density decreased plant-to-plant competition for light, water, and nutrients, which minimized leaf drop caused by shading. Delaying harvest until late flower to green pod maturity stages increased stem yield and maximized total forage yield (Lamb et al., 2003).

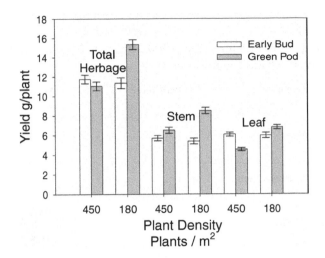

FIGURE 5.1 Mean (±1 SE) for alfalfa total herbage, stem, and leaf yield for each plant density and maturity stage combination.

CHEMICAL COMPOSITION OF ALFALFA

The utility of any biomass crop as a feedstock for ethanol production will depend in large part on its chemical composition, both in terms of the amount of potentially fermentable carbohydrates and the presence of compounds that may limit the yield of these carbohydrates. Current commercial yeast strains only utilize glucose as a substrate for ethanol production. Glucose can be derived from cellulose in the cell walls of biomass species. Therefore cellulose is of greater value than hemicellulose or pectin, polysaccharides composed of numerous sugars other than glucose. However, genetically modified yeast strains and other microorganisms are under study and under development that will use a wider diversity of hexose and pentose sugars. Reduced concentrations of hemicellulose and lignin, a phenolic polymer in the cell wall, would provide benefits to an ethanol conversion system by reducing pretreatment process inputs of heat and acid prior to cellulose addition. Also, reduced lignin content of biomass should result in high concentrations of the cell wall polysaccharides, thereby increasing the potential amount of fermentable sugars. Unfortunately, composition of biomass crops is very diverse and varies due to species, genetics, maturity, and growth environment.

A survey of 190 alfalfa plant introductions in the U.S. germplasm collection found that leaves averaged 283 g crude protein (CP) kg^{-1} dry matter (DM) compared to only 93 g CP kg^{-1} DM in stem material (Jung et al., 1997). In contrast, the neutral detergent fiber (NDF) concentration of stems far exceeded that of leaves (658 and 235 g NDF kg^{-1} DM, respectively). These differences are reflective of the role of stems in providing an upright growth form and supporting the leaf mass. Stems of alfalfa develop extensive xylem tissue (wood) with thick

TABLE 5.1
Composition of Immature (Bud Stage) and Mature (Full Flower) Alfalfa Stem Material

Component	Immature	Mature
	------------------ g kg^{-1} dry matter ------------------	
Protein	127	88
Lipid	9	7
Ash	81	58
Soluble carbohydrates	55	49
Starch	3	2
Cellulose	275	306
Hemicellulose	105	122
Pectin	125	119
Lignin	158	175

Source: Dien, B.S., Jung, H.G., Vogel, K.P., et al., *Biomass Bioenergy*, preprint [submitted].

cell walls comprised of cellulose, hemicellulose, pectin, and lignin (Theander and Westerlund, 1993; Wilson, 1993). Because leaves are the site of most photosynthetic activity in alfalfa, the leaves have high concentrations of enzymes and thin cell walls to facilitate light absorption and gas exchange. Representative composition of alfalfa stem material is shown in Table 5.1. Both leaves and stems have low concentrations of simple sugars and starch (Raguse and Smith, 1966), although alfalfa roots store substantial quantities of starch (150 to 350 g kg^{-1} DM) (Dhont et al., 2002). Lipid content of alfalfa is quite low (~20 g kg^{-1} DM) (Hatfield et al., 2005).

Because alfalfa is indeterminate in its growth habit, the plants increase in size and mass until harvested or a killing frost occurs. Alfalfa leaf mass increases during maturation, but at a lower rate than the increase in stem mass (Sheaffer et al., 2000). This results in a decline in leaf percentage in the total herbage harvested that can range from more than 70% leaf during early vegetative stages to less than 20% leaf when ripe seed is present (Nordkvist and Aman, 1986). During plant maturation, alfalfa leaves change very little in CP or NDF concentration whereas stem CP declines and NDF content increases dramatically (Sheaffer et al., 2000). The reason for the increase in NDF content of alfalfa stems during maturation is the addition of xylem tissue due to cambial activity (Jung and Engels, 2002). This xylem tissue has thick secondary walls and stem xylem accounts for most cell wall material when the crop is harvested.

Cell walls of alfalfa differ from grass cell wall material because of the greater pectin content of alfalfa cell walls. In very immature alfalfa stem internodes that are growing in size, pectins can account for up to 450 g kg^{-1} of the cell wall. Cellulose and hemicellulose contribute 340 and 120 g kg^{-1}, respectively, to the total cell wall, with lignin accounting for the remaining wall material, in such young internodes (Jung and Engels, 2002). At this developmental stage, all of the lignin is localized in the protoxylem vessel cells and no other tissues are lignified. Once alfalfa internodes complete their growth in length, cambium meristematic activity begins to add new xylem fiber and vessel cells that lignify almost immediately. The predominant cell wall component in these tissues is cellulose (400 g kg^{-1} cell wall) with the rest of the cell wall material being equally divided among hemicellulose, pectin, and lignin (Jung and Engels, 2002). Phloem fiber cells also develop thickened secondary cell walls as the plant matures; however, this secondary wall is especially rich in cellulose and does not contain lignin (Engels and Jung, 1998). Lignin is deposited in a unique ring structure in the primary wall region of phloem fiber cells. With the exception of pith paren-chyma cells, all of the other tissues in alfalfa (chlorenchyma, collenchyma, epidermis, cambium, secondary phloem, and protoxylem parenchyma) do not lignify no matter how mature the stem becomes (Engels and Jung, 1998). These tissues retain only primary cell walls that are rich in pectin. The pith parenchyma will ultimately lignify, although with only marginal secondary wall development, but usually pith parenchyma cells senesce, leaving a hollow stem cavity (Jung and Engels, 2002).

The composition of the major cell wall polysaccharides and lignin also change during maturation. Hemicellulose composition shifts from slightly more than 50% xylose residues, with the remainder being primarily to mannose, in very immature elongating stem internodes to 80% xylose residues in very mature internodes (Jung and Engels, 2002). The composition of the pectin fraction shifts less dramatically, with uronic acids increasing from 60% of the pectin to 67% with decreases in galactose and arabinose content, but no change in rhamnose con-centration. The largest shift in cell wall composition due to maturity is in mono-lignol components of lignin. The syringyl-to-guaiacyl ratio increases from 0.29 to 1.01 as alfalfa stem internodes mature (Jung and Engels, 2002).

While maturity is the single most important factor that impacts composition of alfalfa, growth environment causes some additional shifts in composition. Unfortunately these environmental impacts are complex and difficult to predict. In a study by Sanderson and Wedin (1988), alfalfa herbage from a summer regrowth harvest in one year had a substantially higher NDF concentration than observed for that year's spring harvest (538 and 476 g NDF kg^{-1} DM, respec-tively); however, the same plots harvested in the following year showed a small difference between summer and spring harvests (588 and 546 g NDF kg^{-1} DM, respectively). Acid detergent lignin (ADL) concentration of the NDF fraction was greater for summer-harvested alfalfa in both years. During the spring growth period of the second year, air temperatures were warmer and there was less rainfall than in the first year of the study (Sanderson and Wedin, 1988). Vegetatively propagated clones of individual alfalfa plants divergently selected for stem cell wall quality traits showed environmental variability when evaluated over twelve cuttings (two locations, over two years, with three harvests per year). One clone averaged 233 g kg^{-1} for stem Klason lignin concentration but varied in response from 198 to 261 g kg⁻¹ over the environments tested. Another clone selected for stem cellulose concentration ranged from 396 to 467 g kg⁻¹ for the twelve samples (Lamb and Jung, unpublished data).

In the previous study, the impacts of temperature and moisture cannot be evaluated separately. When these two environmental factors have been evaluated independently, the major effect of moisture stress alone appeared to be on amount of cell wall accumulated by alfalfa plants as opposed to changes in cell wall composition. When rainfall was eliminated using a moveable shelter and alfalfa plots were irrigated to three field capacities (65, 88, and 112% saturation), stem cell wall concentration was reduced when the alfalfa was grown under water-deficit conditions (Deetz et al., 1994). Klason lignin concentration of the cell walls was not altered due to water-deficit and concentrations of xylose, galactose, and rhamnose in the cell wall were marginally increased and glucose was decreased, under the 65% field capacity treatment. In contrast to the impact of moisture, temperature was found not to alter cell wall concentration, but did apparently influence cell wall composition. A greenhouse study where alfalfa was grown under adequate moisture conditions indicated that higher temperatures (32°C and 26°C, day and night respectively) resulted in no changes in leaf or stem NDF concentration compared to cooler growth conditions (22°C and 16°C, day and night respectively), but ADL content of the NDF was increased by the higher temperatures (Wilson et al., 1991). However, these temperature effects should be viewed with some caution because both the NDF and ADL concentrations observed for the greenhouse-grown alfalfa in this study were much lower than normally observed for field grown plants.

GENETIC IMPACTS ON COMPOSITION

Genetic differences in chemical composition among alfalfa plant introductions, varieties, and individual genotypes have been reported. Leaf and stem CP differed among a group of 61 plant introductions, although the ranges were small, from 272 to 295 and 88 to 99 g CP kg⁻¹ DM, respectively (Jung et al., 1997). Leaf NDF concentration (235 g kg⁻¹ DM) did not differ significantly among these plant introductions, but stem NDF ranged from 636 to 670 g NDF kg⁻¹ DM. Similar variation was observed among a group of five commercial alfalfa varieties with CP and NDF differences being noted for leaves and stems, as well as whole herbage (Sheaffer et al., 2000). Differences in stem cell wall concentration and composition were observed among a set of four alfalfa genotypes selected for divergence in whole herbage ADL and in vitro ruminal DM disappearance (IVDMD) (Jung et al., 1994) and a group of three genotypes selected for divergent IVDMD (Jung and Engels, 2002). More recently, alfalfa genotypes selected for divergent cell wall Klason lignin, cellulose, and xylan were shown to differ genetically for these cell wall components when grown across a series of environments (Lamb and Jung, 2004). While the reported genetic variation among alfalfa germplasm sources is not large, the potential for modifying cell wall composition has not been seriously explored, because recurrent selection for these traits has not been done.

Significant genotype x environment (G × E) interactions have generally not been observed for chemical composition of alfalfa varieties. Among 61 plant introductions, no measures of cell wall concentration or composition were found to have significant G × E interactions for leaf or stem material (Jung et al., 1997). Only differences in magnitude, not rank, for composition due to G × E interactions were noted by Sheaffer et al. (2000) among five alfalfa varieties. These results mirror the conclusion of Buxton and Casler (1993) that forage quality traits generally have small G × E interaction effects compared to the impact on yield. However, in recent work with alfalfa clones selected for specific cell wall traits, it was found that G × E interactions were significant among plants selected for low and high pectin and xylan concentrations, whereas no G × E interactions were noted among clones selected for Klason lignin or cellulose (Lamb and Jung, 2004).

ALFALFA LEAF MEAL

Because alfalfa leaves contain approximately 300 g CP kg^{-1} DM, this portion of the crop has greater value as an animal feedstuff than for conversion to ethanol. Based simply on its protein concentration, alfalfa leaf meal was estimated to have a value of $138 Mg^{-1} (Linn and Jung, unpublished). This price far exceeds the target feedstock value of $33 Mg^{-1} assumed in a functioning corn stover-to-ethanol production system (Aden et al., 2002). In an extensive series of studies involving lactating dairy cows and fattening beef cattle, alfalfa leaf meal was shown to be an acceptable protein feed supplement in place of soybean meal (DiCostanzo et al., 1999). Besides providing protein for beef steer growth, alfalfa leaf meal also reduced the incidence of liver abscesses at slaughter, thereby increasing the market value of the cattle. Furthermore, alfalfa leaf meal could replace alfalfa hay in the diet of lactating dairy cows as a source of both protein and fiber to support normal milk production (Akayezu et al., 1997). Suckling beef calves actually gained weight more rapidly when fed alfalfa leaf meal in a supplemental creep feed than observed with a soybean meal-based supplement (DiCostanzo et al., 1999). From these results, it is clear that alfalfa leaf meal could provide a valuable coproduct for an alfalfa-to-ethanol production system.

PROTEIN AND FIBER SEPARATION

Two methods have been developed for capturing the protein-rich fraction from alfalfa and separating it from the more fiber-rich fraction. From whole field-dried plant material, leaves can be separated from denser stems using shaking screens (Arinze et al., 2003; Downing et al., 2005). Fresh material can be dried using a rotary drum drier and leaves separated aerodynamically due to their lower mass and faster drying time than that of stems (Arinz et al., 2003). Wet fractionation involves mechanical maceration of fresh total herbage followed by the expression of protein-rich juice (Jorgensen and Koegel, 1988; Koegel and Straub, 1996). Approximately 20–30% of the herbage DM can be captured in the juice (Koegel and Straub, 1996). The proportion of DM that was captured in the juice was shown to decrease with increasing maturity of the herbage (Koegel and Straub, 1996). The juice contains both particulate and soluble proteins. The soluble proteins, which may have greater value, can be separated from particulate proteins by heating and centrifugation (Jorgensen and Koegel, 1988). Wet fractionation has been used successfully in small-scale experiments (see "Pretreatment of Alfalfa Fiber" below) to refine alfalfa into a high-value protein fraction and a fiber fraction that was further refined and fermented to produce ethanol (Koegel et al., 1999; Sreenath et al., 2001), lactic acid (Koegel et al., 1999), and wood adhesive (Weimer et al., 2005). Fiber can also be processed into animal feed. The deproteinized juice is a source for extracting xanthophyll and can also be used as a fertilizer (Koegel and Straub, 1996). Wet fractionation has the advantage of minimizing leaf loss and is less weather dependent than field drying. Dried material has the advantage of being lighter to transport and is easily stored for later processing and refining. The nature of the protein product will clearly impact the method of herbage harvest and processing.

In addition to protein, alfalfa also contains numerous secondary metabolites that are of interest in human nutrition and food production. In particular, alfalfa is a rich source of flavonoid antioxidants and phytoestrogens including luteolin, coumestrol, and apigenin (Hwang et al., 2001; Stochmal et al., 2001) that have possible health-promoting activities. Alfalfa foliage also contains high amounts of xanthophylls, which are added to chicken feed to pigment egg yolks and broiler skin (Koegel and Straub 1996). Thus, in a biorefinery model for alfalfa processing, ethanol would be one of several products produced with the protein component possibly the more valuable and economically important product.

PRETREATMENT OF ALFALFA FIBER

Ethanol production depends on fermentation of simple sugars by microorganisms. The yield of potentially fermentable sugars from the conversion process is the critical response variable in assessing the value of alfalfa as an ethanol production feedstock. Potentially fermentable sugar yield is a function of both carbohydrate composition and concentration (discussed earlier), and the efficiency with which the cell wall polysaccharides are converted to simple sugars through processing. The results of two pretreatment methods have been reported previously. Ferrer et al. (2002) described parameters of ammonia processing of whole dried alfalfa hay that influenced the susceptibility of the fiber to subsequent enzymatic hydrolysis. The ammonia loading, moisture, time and temperature of treatment were varied and then the treated material digested with a mixture of cellulase, cellobiase, and xylanase. Conditions of 2 g ammonia g^{-1} DM, with 30% moisture and processing at 85°C for five minutes was shown to convert 76% of the theoretical yield of reducing sugars in the fiber. Approximately 200 mg sugars g^{-1} DM was obtained (Ferrer et al., 2002); however, the yield of ethanol produced from this material remains to be determined.

Liquid hot water (LHW) pretreatments of the fiber fraction obtained after wet fractionation of alfalfa have been optimized for maximum sugar conversion (Sreenath et al., 1999) and ethanol production (Sreenath et al., 2001). The LHW pretreatment was found to solubilize hemicellulose, and the resulting extract contained significant amounts of acetic acid and formic acid (Sreenath et al., 1999). The remaining fiber fraction (raffinate) when treated with cellulase released 59 g of reducing sugars from 100 g of substrate. Addition of dilute acid (0.07% sulfuric acid) to the LHW decreased the amount of reducing sugars released by cellulase treatment to 24 g 100 g^{-1} substrate (Sreenath et al., 1999). Fermentation of the raffinate fraction after LHW pretreatment was tested with two strains of *Candida shehatae* in a simultaneous saccharification and fermentation (SSF) process as well as a separate hydrolysis and fermentation (SHF) process (Sreenath et al., 2001). The yield of ethanol was 0.45 g ethanol g^{-1} sugar with SSF and 0.47g ethanol g^{-1} sugar with SHF. The extract from the LHW pretreatment was also used in fermentation experiments and was poorly fermented, most likely due to the presence of organic acids. Addition of dilute acid to the LHW treatment resulted in fractions that were poorly fermented. Although untreated fiber substrate was shown to yield 51 g reducing sugars from 100 g of substrate (Sreenath et al., 1999), the yield of ethanol by SHF and SSF was 0.25 and 0.16 g ethanol g^{-1} sugar, respectively (Sreenath et al., 2001). These experiments demonstrate the impact of pretreatment on saccharification and ethanol production as well as the requirement to optimize processes for each lignocellulosic feedstock.

CONVERSION RESPONSE AFTER DILUTE ACID PRETREATMENT

For the purposes of this chapter, the high temperature, dilute acid pretreatment and subsequent enzymatic saccharification method will be examined in more detail as a conversion technology for ethanol production from alfalfa stem fractions. The high temperature, dilute acid pretreatment is designed to remove noncellulosic cell wall polysaccharides and lignin, because these constituents will interfere with the cellulase enzyme cocktails used for hydrolysis of the cellulose. One design goal of this pretreatment is to reduce the pH of the feedstock reaction mixture to 1.3–1.5 prior to heating (National Renewal Energy Laboratory, Golden, CO; Laboratory Analytical Procedure-007, May 17, 1995). The amount of sulfuric acid required to reach this pH target for alfalfa stems was 8.1 mmol g^{-1} biomass DM in a 1% solids slurry, compared to 6.4 mmol for switchgrass and corn stover (Jung, unpublished). Maturity of alfalfa stems and switchgrass did not influence the acid requirement. Dien et al. (2005) observed that the sulfuric acid loading required to maximize release of nonglucose sugars from alfalfa stems when heated at 121°C for 1 h was 2.5% (wt/vol), whereas 1.5% was sufficient for switchgrass. The higher acid requirement for alfalfa stems is most likely due to the greater pectin content of alfalfa cell

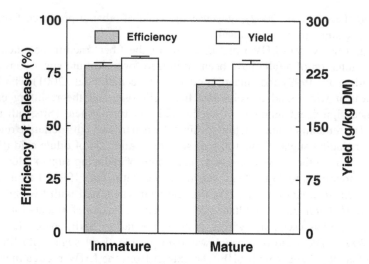

FIGURE 5.2 Efficiency of conversion and total yield of glucose from alfalfa stems when pretreated with dilute sulfuric acid at 150°C and subsequently saccharified using cellulase. (Dien et al., 2005).

walls compared to grasses; however, the hemicellulose content is lower and lignin content is similar in alfalfa stems compared to the grasses (Dien et al., 2005). Torget et al. (1990, 1992) also observed that legume feedstocks are more recalcitrant to acid pretreatment than grasses.

The efficiency of glucose release by acid pretreatment, followed by enzymatic saccharification from cell wall polysaccharides (cellulose and xyloglucans), declined as alfalfa stems became more mature (Figure 5.2). While efficiency of glucose conversion declined with maturity, the total yield of glucose was not altered (Figure 5.2), because cellulose content increased in more mature alfalfa stems. Similar declines in efficiency with maturity were observed for switchgrass and reed canary grass (Dien et al., 2005), but the efficiency of glucose release from the grasses was greater than from alfalfa stems. This may reflect the higher concentration of lignin in the alfalfa stems because across all three species, efficiency of glucose release was negatively correlated with lignin content of the feedstock. Increasing the temperature of the acid pretreatment resulted in improved efficiency of glucose release from alfalfa stems (Dien, personal communication). While efficiency of glucose release was lower for alfalfa than for grasses, total yield of glucose was very similar between the feedstocks. This again reflects the interaction of efficiency with glucose content of the feedstocks.

ALFALFA BIOTECHNOLOGY AND GENOMICS

An additional characteristic of alfalfa that makes it attractive for biorefinement is that it is amenable to genetic transformation. Rapid and efficient methods for transformation using *Agrobacterium tumefaciens* have been developed and gene

TABLE 5.2
Transgenic Alfalfa Producing Commercial Enzymes and Polymers

Enzyme	Gene	Source	Amount of Product	Citation
Phytase	phyA	*Aspergillus ficuum*	0.85–1.8% of soluble protein	Austin-Phillips and Ziegelhoffer, 2001 Ullah et al., 2002
Manganese-dependent lignin peroxidase	Mn-P	*Phanerochaete chrysoporium*	0.01–0.5% of soluble protein	Austin et al., 1995
α-amylase	α-amylase	*Bacillus licheniformis*	0.001–0.01% of soluble protein	Austin et al., 1995
Endo-glucanase	E2	*Thermomonospora fusca*	0.01% of soluble protein	Ziegelhoffer et al., 1999
Cellobiohydrolase	E3	*Thermomonospora fusca*	0.001–0.002% of soluble protein	Ziegelhoffer et al., 1999
β-ketothiolase	phbA	*Ralstonia eutropha*	0.025–1.8 g PHB/kg dry leaves	Saruul et al., 2002
Acetoacetyl-CoA reductase	phbB			
PHB synthase	phbC			

promoters identified for high constitutive expression and for tissue-specific expression (reviewed by Samac and Temple, 2004; Somers et al., 2003). Transformation has been used to alter alfalfa for production of valuable coproducts (Table 5.2) and for improving digestion of alfalfa fiber. Transgenic alfalfa has been shown to be capable of producing high levels of phytase (Austin-Phillips and Ziegelhoffer, 2001; Ullah et al., 2002), a feed enzyme that degrades phytic acid and makes phosphorus in vegetable feeds available to monogastric animals such as swine. Adding phytase to feeds reduces the need to add supplemental phosphorus to feed and reduces the amount of phosphorus excreted by animals. In field studies, juice from wet-fractionated alfalfa plants contained 1–1.5% phytase. Phytase activity in juice was stable over two weeks at a temperature of 37°C. Activity is also stable in dried leaf meal. Both juice and dried leaf meal added to feed were as effective in feeding trials as phytase from microbial sources. The value of the enzyme and xanthophyll in the juice was estimated at $1900/acre (Austin-Phillips and Ziegelhoffer, 2001). A wide range of feed enzymes is used to enhance digestion of feed and improve animal performance. Use of feed enzymes in monogastric and ruminant animals in expected to increase worldwide (Sheppy, 2001). Production of feed enzymes in transgenic plants, particularly in plants used as animal feed, would be an opportunity to increase feed utilization as well as value of the feed.

Transgenic alfalfa has also been used to produce several industrial enzymes. A manganese-dependent lignin peroxidase, which can be used for lignin degradation and biopulping in the manufacture of paper, was expressed in alfalfa. However, high levels of production of this enzyme appeared to be detrimental to plants (Austin et al., 1995). In the same study, α-amylase was produced at a level of approximately 0.01% of soluble protein without having a negative effect on plant development. Two cellulases, an endogluconase and a cellobiohydrolase, have been expressed at low levels in alfalfa (Ziegelhoffer et al., 1999). These enzymes were stable in dried leaf meal. Expression of cellulose degrading enzymes in biomass plants is one strategy to decrease the costs of saccharification that precedes ethanol fermentation. Alfalfa plants have also been shown to be an excellent "factory" for the production of chitinase (Samac et al., 2004). Chitin, found in shells of crustaceans, is the second most abundant carbohydrate after cellulose, and a potential feedstock in a biorefinery.

In addition to production of proteins, the use of transgenic alfalfa to produce other industrial feed stocks has been explored. Polyhydroxyalkanoates (PHAs) are produced by many species of bacteria and some PHA polymers are commercially valuable as biodegradable plastics. PHA synthesis in plants is seen as a more economically viable means of producing large quantities of these polymers (Poirier, 1999; Slater et al., 1999). Alfalfa was engineered to constitutively express three bacterial genes for the production of poly-β-hydroxybutyrate (PHB) (Saruul et al., 2002). Granules of PHB were shown to accumulate in chloroplasts without any negative impact on plant growth. Yield of PHB by chemical extraction was relatively low (1.8 g kg^{-1} DM), but may be improved by optimizing extraction methods or by utilizing stronger gene promoters.

A major limitation to use of biomass in the production of ethanol is the recalcitrance of the material to saccharification. Cross-linking of lignin with cell-wall polysaccharides interferes with enzymatic degradation of cellulose and can severely limit the conversion of herbaceous plant material into ethanol. Lignin in alfalfa stems also limits digestion of feed by ruminant animals. In experiments aimed at increasing feed digestion by ruminants, transgenic alfalfa was produced that had decreased expression of caffeoyl coenzyme A 3-*O*-methyltransferase, an enzyme involved in synthesis of lignin precursors. These plants were shown to have approximately 20% less lignin and 10% additional cellulose than the controls (Marita et al., 2003). The rate of digestion of the transgenic material was determined by *in vitro* rumen digestibility assays. In the transgenic material, a 2.8–6.0% increase in the rate of digestion was observed (Guo et al., 2001). This material could have a very significant impact on both animal nutrition and alfalfa biorefining. Casler and Vogel (1999) determined that a 1% increase in forage digestibility would lead to a 3.2% increase in average daily live-weight gain by beef steers. Although this material has not yet been tested with different pretreatment methods or used in saccharification or fermentation studies, based on chemical analyses, it may also have improved qualities as a feedstock for bioethanol production.

During the past several years, barrel medic (*Medicago truncatula*) has been the object of a broad range of research efforts worldwide. This annual plant, which is closely related to alfalfa, is a model plant for study of plant-microbe interactions and plant development (Cook, 1999). Chromosome mapping has shown that there is a high degree of gene synteny between the two species as well as a high degree of DNA sequence homology (Choi et al., 2004). Numerous genomic tools have been developed for *M. truncatula* including isolation of over 189,000 expressed sequence tags (ESTs), identification and sequencing of more than 36,000 unique genes (http://www.tigr.org/tigr-scripts/tgi/T_index.cgi?species=medicago), extensive genetic and physical mapping (Choi et al. 2004), development of microarrays for transcript profiling, and a genome sequencing project is currently underway (http://www.medicago.org). In particular, microarrays are valuable tools for identifying genes involved in important agricultural processes as they enable researchers to measure expression of thousands of genes simultaneously. More than 100 genes are involved in cell-wall biosynthesis in plants and little is known about regulation of their expression. EST resources may be useful both as markers for selecting plants with favorable characteristics in bioconversion and in modifying gene expression in transgenic plants for enhancing the efficiency of ethanol production or enhancing yields of valuable coproducts.

CONCLUSIONS

Although commercial biorefining of alfalfa remains undeveloped, alfalfa has tremendous potential as a feedstock for production of ethanol and other products. Alfalfa is widely adapted and produces large amounts of biomass over the course of four or more years. The production costs of alfalfa are low and cultivation of the crop has numerous environmental benefits. Importantly, alfalfa leaves contain the majority of the protein in the plant and are easily separated from stems through processing. Leaf meal is a valuable coproduct in its own right as animal feed, as well as a potential source for human nutritional supplements and products derived from transgene expression. The stem fraction of alfalfa is rich in cell wall polysaccharides that can be used as a source of fermentable sugars to produce ethanol and other bioproducts. A biomass-type of alfalfa is being developed that is more upright in growth habit and performs well in a reduced frequency harvest management system, maximizing the yield of both leaf and stem fractions while lowering production costs. Incorporation of enhanced compositional traits such as more cellulose, less lignin and valuable transgenic protein products into this alfalfa biomass type through traditional breeding and using the tools of biotechnology will add to the value alfalfa brings to biofuels and bioproduct systems.

ACKNOWLEDGMENTS

The authors thank members of the USDA-ARS-Plant Science Research Unit (Drs. Carroll P. Vance, Michael P. Russelle, and John Gronwald) and Dr. Sandra Austin-Phillips (University of Wisconsin-Madison) for stimulating conversations, insights, and for providing literature citations.

REFERENCES

Aden, A., Ruth, M., Ibsen, K., Jechura, J., Neeves, K., Sheehan, J., Wallace, B., Montague, L., Slayton, A. and Lukas, J., *Lignocellulosic Biomass to Ethanol Process Design and Economics Utilizing Co-current Dilute Acid Prehydrolysis and Enzymatic Hydrolysis for Corn Stover*, NREL/TP-510-32438, National Renewable Energy Laboratory, Golden, CO, 2002, pp. 1–88.

Akayezu, J.M., Jorgensen, M., Linn, J.G. and Jung, H.G., Effects of Substituting Alfalfa Leaf Meal for Alfalfa Hay in Diets of Early Lactation Cows, *J. Dairy Sci.*, 80, 184, 1997.

Arinze, E.A., Schoenau, G.J., Sokhansanj, S. and Adapa, P., Aerodynamic Separation and Fractional Drying of Alfalfa Leaves and Stems—A Review and New Concept, *Drying Technol.*, 21, 1669–1698, 2003.

Austin, S., Bingham, E.T., Mathews, D.E., Hahan, M.N., Will, J. and Burgess, R.R., Production and Field Performance of Transgenic Alfalfa (*Medicago sativa* L.) Expressing Alpha-Amylase and Manganese-Dependent Lignin Peroxidase, *Euphytica*, 4, 1–13, 1995.

Austin-Phillips, S. and Ziegelhoffer, T., The Production of Value-Added Proteins in Transgenic Alfalfa,. In *Molecular Breeding of Forage Crops*, Spangenberg, G., Ed., Kluwer Academic Publishers, Dordrecht, The Netherlands 2001, pp. 285–301.

Barnes, D.K., Bingham, E.T., Murphy, R.P., Hunt, O.J., Beard, D.F., Skrdla, W.H. and Teuber, L.R., *Alfalfa Germplasm in the United States: Genetic Vulnerability, Use, Improvement, and Maintenance,*. U.S. Department of Agriculture, Tech. Bull. 1571, Washington, DC, 1977, pp. 1–21.

Brummer, E.C., Capturing Heterosis in Forage Crop Cultivar Development,. *Crop Sci.*, 39, 943–954, 1999.

Buxton, D.R. and Casler, M.D., Environmental and Genetic Effects on Cell Wall Composition and Digestibility, In *Forage Cell Wall Structure and Digestibility*, Jung, H.G., Buxton, D.R., Hatfield, R.D. and Ralph, J. Eds., American Society of Agronomy-Crop Science Society of American-Soil Science Society of America Publishers, Madison, WI, 1993, pp. 685–714.

Casler, M.D. and Vogel, K.P., Accomplishments and Impact from Breeding for Increased Forage Nutritional Value, *Crop Sci.*, 39, 12–20, 1999.

Choi, H.K., Kim, D., Uhm, T., Limpens, E., Lim, H., Mun, J.H., Kalo, P., Penmetsa, R.V., Seres, A., Kulikova, O., Roe, B.A., Bisseling, T., Kiss, G.B. and Cook, D.R., A Sequence-Based Genetic Map of *Medicago truncatula* and Comparison of Marker Colinearity with *M. sativa*, *Genetics*, 166, 1463–1502, 2004.

Cook, D.R., *Medicago truncatula*: A model in the making,. *Curr. Opin. Plant Biol.*, 2, 301–304, 1999.

Cowett, E.R. and Sprague, M A., Factors Affecting Tillering in Alfalfa, *Agron. J.*, 54, 294–297, 1962.

Dale, B.E., Biomass Refining: Protein and Ethanol from Alfalfa, *Ind. Eng. Chem. Prod. Res. Dev.*, 22, 466–472, 1983.

Deetz, D.A., Jung, H.G. and Buxton, D.R., Water-Deficit Effects on Cell-Wall Composition and In Vitro Degradability of Structural Polysaccharides from Alfalfa Stems, *Crop Sci.*, 36, 383–388, 1994.

Dhont, C., Castonguay, Y., Nadeau, P., Belanger, G. and Chalifour, F.-P., Alfalfa Root Carbohydrates and Regrowth Potential in Response to Fall Harvests, *Crop Sci.*, 42, 754–765, 2002.

Dien, B.S., Jung, H.G., Vogel, K.P., Casler, M.D., Lamb, J.F.S., Weimer, P.J., Iten, L., Sarath, G. and Mitchell, R., Chemical Composition and Response to Dilute-Acid Pretreatment and Enzymatic Saccharification of Alfalfa, Reed Canarygrass, and Switchgrass, *Biomass Bioenergy*, preprint [submitted].

DiCostanzo, A., Zehnder, C.M., Akayezu, J.M., Jorgensen, M.A, Cassidy, J.M., Allen, D.M., Standorf, D.G.; Linn, J.G., Jung, H., Smith, L.J., Lamb, G.C., Johnson, D., Chester-Jones, H. and Robinson, G., *Use of Alfalfa Leaf Meal in Ruminant Diets*. 60th Minnesota Nutrition Conference, Bloomington, MN, Sept. 20–22, 1999, pp. 64–75.

Downing, M., Volk, T.A. and Schmidt, D.A., Development of New Generation Cooperatives in Agriculture for Renewable Energy Research, Development, and Demonstration Projects, *Biomass Bioenergy*, 28, 425–434, 2004.

Engels, F.M. and Jung, H.G., Alfalfa Stem Tissues: Cell-Wall Development and Lignification, *Ann. Bot.*, 82, 561–568, 1998.

Ferrer, A., Byers, F.M., Sulbaran-de-Ferrer, B., Dale, B.E. and Aiello, C., Optimizing Ammonia Processing Conditions to Enhance Susceptibility of Legumes to Fiber Hydrolysis-Alfalfa, *App. Biochem. Biotech.*, 98, 123–143, 2002.

Guo, D., Chen, F., Wheeler, J., Winder, J., Selman, S., Peterson, M. and Dixon, R.A. Improvement of In-Rumen Digestibility of Alfalfa Forage by Genetic Manipulation of Lignin O-Methyltransferases, *Transgenic Res.*, 10, 457–464, 2001.

Hansen, L.H. and Krueger, C.R., Effect of Establishment Method, Variety, and Seeding Rate on the Production and Quality of Alfalfa Under Dryland and Irrigation, *Agron. J.*, 65, 755–759, 1973.

Huggins, D.R., Randall, G.W. and Russelle, M.P., Subsurface Drain Losses of Water and Nitrate Following Conversion of Perennials to Row Crops, *Agronomy J.*, 93, 477–486, 2001.

Hwang, J.L., Hodis, H.N. and Sevanian, A., Soy and Alfalfa Phytoestrogen Extracts Become Potent Low-Density Lipoporotein Antioxidants in the Presence of Acerola Cherry Extract, *J. Agric. Food Chem.*, 49, 308–314, 2001.

Johnson, L.D., Marquez-Ortiz, J.J., Barnes, D.K. and Lamb, J.F.S., Inheritance of Root Traits in Alfalfa, *Crop Sci.*, 36, 1482–1487, 1996.

Jorgensen, N.A. and Koegel, R.G., Wet Fractionation Processes and Products, In *Alfalfa and Alfalfa Improvement*, Hanson, A.A., Barnes, D.K. and Hill, R.R, Eds, American Society of Agronomy-Crop Science Society of American-Soil Science Society of America Publishers, Madison, WI, 1988, pp. 553–566.

Jung, H.G. and Engels, F.M., Alfalfa Stem Tissues: Cell-Wall Deposition, Composition, and Degradability, *Crop Sci.*, 42, 524–534, 2002.

96   Alcoholic Fuels

Jung, H.G., Smith, R.R. and Endres, C.S., Cell-Wall Composition and Degradability of Stem Tissue from Lucerne Divergently Selected for Lignin and *In Vitro* Dry Matter Disappearance, *Grass Forage Sci.*, 49, 1–10, 1994.

Jung, H.G., Sheaffer, C.C., Barnes, D.K. and Halgerson, J.L., Forage Quality Variation in the U.S. Alfalfa Core Collection, *Crop Sci.* 37, 1361–1366, 1997.

Kiesselbach, T.A., Russel, J.C. and Anderson, A., The Significance of Subsoil Moisture in Alfalfa Production. *J. Am. Soc. Agron.*, 21, 241–268, 1929.

Kim, S. and Dale, B.D., Cumulative Energy and Global Warming Impact from the Production of Biomass for Biobased Products, *J. Indust. Ecol.*, 7, 147–162, 2004.

Koegel, R.G., Sreenath, H.K. and Straub, R.J., Alfalfa Fiber as a Feedstock for Ethanol and Organic Acids, *App. Biochem. Biotech.*, 77–79, 105–115, 1999.

Koegel, R.G. and Straub, R.J., Fractionaltion of Alfalfa for Food, Feed, Biomass, and Enzymes, *Trans. Amer. Soc. Ag. Eng.*, 39, 769–774, 1996.

Lamb, J.F.S. and Jung, H.G., *Environmental Stability of Stem Quality Traits in Alfalfa*, American Society of Agronomy, Crop Science Society of American, Soil Science Society of America Annual Meetings, Seattle, WA, Oct 31–Nov 4, 2004, Abstr. 3885.

Lamb, J.F.S., Sheaffer, C.C. and Samac, D.A., Population Density and Harvest Maturity Effects on Leaf and Stem Yield in Alfalfa, *Agron. J.*, 295, 635–641, 2003.

Marita, J.M., Ralph, J., Hatfield, R.D., Guo, D., Chen, F. and Dixon, R.A., Structural and Compositional Modifications in Lignin of Transgenic Alfalfa Down-Regulated in Caffeic Acid 3-*O*-Methyltransferase and Caffeoyl Coenzyme A 3-*O*-Methyltransferase, *Phytochemistry*, 62, 53–65, 2003.

Marquez-Ortiz, J.J., Lamb, J.F.S., Johnson, L.D., Barnes, D.K. and Stucker, R.E., Heritability of Crown Traits in Alfalfa, *Crop Sci.*, 39, 38–43, 1999.

Nordkvist, E. and Aman, P., Changes During Growth in Anatomical and Chemical Composition and *In-Vitro* Degradability of Lucerne, *J. Sci. Food Agric.*, 37, 1–7, 1986.

Poirier, Y., Production of New Polymeric Compounds in Plants., *Curr. Op. Biotech.*, 10, 181–185, 1999.

Raguse, C.A. and Smith, D., Some Nonstructural Carbohydrates in Forage Legume Herbage, *J. Agric. Food Chem.*, 14, 423–426, 1966.

Russelle, M.P., Alfalfa, *Amer. Sci.*, 89, 252–261, 2001.

Rumbaugh, M.D., Effects of Population Density on Some Components of Yield of Alfalfa, *Crop Sci.*, 3, 423–424, 1963.

Samac, D.A. and Temple, S.J., Development and Utilization of Transformation in *Medicago* Species,. In *Genetically Modified Crops: Their Development, Uses, and Risks*, Liang, G.H. and Skinner, D.Z., Eds., Food Products Press, Binghamton, NY, 2004, pp. 165–202.

Samac, D.A., Tesfaye, M., Dornbusch, M., Saruul, P. and Temple, S.J., A Comparison of Constitutive Promoters for Expression of Transgenes in Alfalfa (*Medicago sativa*), *Transgenic Res.*, 13, 349–361, 2004.

Sanderson, M.A. and Wedin, W.F., Cell Wall Composition of Alfalfa Stems at Similar Morphological Stages and Chronological Age During Spring Growth and Summer Regrowth, *Crop Sci.*, 28, 342–347, 1988.

Saruul, P., Srienc, F., Somers, D.A. and Samac, D.A., Production of a Biodegradable Plastic Polymer, Poly-β-Hydroxybutyrate, in Transgenic Alfalfa, *Crop Sci.*, 42, 919–927, 2002.

Sheaffer, C.C., Martin, N.P., Lamb, J.F.S., Cuomo, G.R., Jewett, J.G. and Quering, S.R., Leaf and Stem Properties of Alfalfa Entries, *Agron. J.*, 92, 733–739, 2000.

Sheppy, C., The Current Feed Enzyme Market and Likely Trends, In *Enzymes in Farm Animal Nutrition*, Bedford, M.R. and Partridge, G.G., Eds., CABI Publishing, Wallingford, CT, 2001, pp. 273–298.

Slater, S., Mitsky, T.A., Houmiel, K.L., Hao, M., Reiser, S.E., Taylor, N.B., Tran, M., Valentin, H.E., Rodriguez, D.J., Stone, D.A., Pagette, S.R., Kishore, G. and Gruys, K.J., Metabolic Engineering of *Arabidopsis* and *Brassica* for Poly(3-Hydroxybutyrate-*Co*-3-Hydroxyvalerate) Copolymer Production, *Nat. Biotech.*, 17, 1011–1016, 1999.

Somers, D.A., Samac, D.A. and Olhoft, P.M., Recent Advances in Legume Transformation, *Plant Physiol.*, 131, 892–899, 2003.

Sreenath, H.K., Koegel, R.G., Moldes, A.B., Jeffries, T.W. and Straub, R.J., Enzymic Saccharification of Alfalfa Fibre After Liquid Hot Water Pretreatment, *Process Biochem.*, 35, 33–41, 1999.

Sreenath, H.K., Koegel, R.G., Moldes, A.B., Jeffries, T.W. and Straub, R.J., Ethanol Production from Alfalfa Fiber Fractions by Saccharification and Fermentation, *Process Biochem.*, 36, 1199–1204, 2001.

Stochmal, A., Piacente, S., Pizza, C., DeRiccardis, F., Leitz, R. and Oleszek, W., Alfalfa (*Medicago sativa* L.) Flavonoids: 1. Apigenin and Luteolin Glycosides From Aerial Parts, *J. Agric. Food Chem.*, 49, 753–758, 2001.

Theander, O. and Westerlund, E., Quantitative Analysis of Cell Wall Components, In *Forage Cell Wall Structure and Digestibility*, Jung, H.G., Buxton, D.R., Hatfield, R.D. and Ralph, J., Eds., American Society of Agronomy-Crop Science Society of American-Soil Science Society of America Publishers, Madison, WI, 1993, pp. 83–104.

Torget, R., Himmel, M. and Grohmann, K., Dilute-Acid Pretreatment of Two Short-Rotation Herbaceous Crops: Scientific Note, *Appl. Biochem. Biotechnol.*, 28, 29, 115–123, 1992.

Torget, R., Werdene, P., Himmel, M. and Grohmann, K., Dilute Acid Pretreatment of Short Rotation Woody and Herbaceous Crops, *Appl. Biochem. Biotechnol.*, 24, 25, 115–126, 1990.

Ullah, A.H., Sethumadhavan, K., Mullaney, E.J., Ziegelhoffer, T. and Austin-Phillips, S., Cloned and Expressed Fungal *phyA* Gene in Alfalfa Produces a Stable Phytase, *Biochem. Biophys. Res. Commun.*, 290, 1343–1348, 2002.

USDA-NASS, Agricultural Statistics 2004, http://www.usda.gov/nass/pubs/agr04/04_ch6.pdf

Volenec, J.J., Cherney, J.H. and Johnson, K.D., Yield Components, Plant Morphology, and Forage Quality of Alfalfa as Influenced by Plant Population, *Crop Sci.*, 27, 321–326, 1987.

Weimer, P.J., Koegel, R.G., Lorenz, L.F., Frihart, C.R. and Kenealy, W.R., Wood adhesives prepared from lucerne fiber fermentation residues of *Ruminococcus albus* and *Clostridium thermocellum*, *Appl Microbiol. Biotechnol.*, 66, 635–640, 2005.

Wilson, J.R., Organization of Forage Plant Tissues, In *Forage Cell Wall Structure and Digestibility*, Jung, H.G., Buxton, D.R., Hatfield, R.D. and Ralph, J., Eds., American Society of Agronomy-Crop Science Society of American-Soil Science Society of America Publishers, Madison, WI, 1993, pp. 1–32.

Wilson, J.R., Deinum, B. and Engels, F.M., Temperature Effects on Anatomy and Digestibility of Leaf and Stem of Tropical and Temperate Forage Species, *Neth. J. Agric. Sci.*, 39, 31–48, 1991.

Ziegelhoffer, T., Will, J. and Austin-Phillips, S., Expression of Bacterial Cellulase Genes in Transgenic Alfalfa (*Medicago sativa* L.), Potato (*Solanum tuberosum* L.) and Tobacco (*Nicotiana tabacum* L.), *Mol. Breed.*, 5, 309–318, 1999.

6 Production of Butanol from Corn

*Thaddeus C. Ezeji,[1] Nasib Qureshi,[2]
Patrick Karcher,[1] and Hans P. Blaschek[1]*

[1]University of Illinois, Biotechnology & Bioengineering Group, Department of Food Science & Human Nutrition, Urbana

[2]United States Department of Agriculture, National Center for Agricultural Utilization Research, Fermentation/Biotechnology, Peoria, Illinois

CONTENTS

Abstract The last few decades have witnessed dramatic improvements made in the production of fuels and chemicals from biomass and fermentation derived butanol production from corn is no exception. The art of producing butanol from corn that existed during World Wars I and II is no longer seen as an art but rather as science. Recent developments have brought, once again, the forgotten acetone butanol ethanol (ABE) fermentation from corn closer to commercialization. Superior strains have been developed, along with state-of-the-art upstream, downstream, and fermentation technologies. Butanol can be produced not only from corn starch as was done decades ago, but also from corn coproducts such as corn fiber and corn steep liquor (CSL) as a nutrient supplement. These additional substrates add to the improved yield and superior economics of the butanol process. Downstream processing technologies have enabled the use of concentrated sugar solutions to be fermented, thereby resulting in improved process efficiencies. Application of fed-batch fermentation in combination with *in situ*/inline product recovery by gas stripping and pervaporation is seen as a superior technology for scale-up of butanol production. Similarly, continuous fermentations (immobilized cell and cell recycle) have resulted in dramatic improvement in reactor productivities. This chapter details all the above developments that have been made for production of butanol from corn. As of today, butanol production from corn is competitive with petrochemically produced butanol.

INTRODUCTION

Butanol is a four-carbon alcohol, a clear neutral liquid with a strong characteristic odor. It is miscible with most solvents (alcohols, ether, aldehydes, ketones, and aliphatic and aromatic hydrocarbons), is sparingly soluble in water (water solubility 6.3%) and is a highly refractive compound. Currently, butanol is produced chemically by either the oxo process starting from propylene (with H_2 and CO over rhodium catalyst) or the aldol process starting from acetaldehyde (Sherman, 1979). Butanol is also produced by fermentation of corn and corn-milling by-products. Butanol is a chemical that has excellent fuel characteristics. It contains approximately 22% oxygen, which when used as a fuel extender will result in more complete fuel combustion. Use of butanol as fuel will contribute to clean air by reducing smog-creating compounds, harmful emissions (carbon monoxide) and unburned hydrocarbons in the tail pipe exhaust. Butanol has research and motor octane numbers of 113 and 94, compared to 111 and 92 for ethanol (Ladisch, 1991). Some of the advantages of butanol as a fuel have been reported previously (Ladisch, 1991).

Butanol production by fermentation dates back to Louis Pasteur (1861) who discovered that bacteria can produce butanol. In 1912, Chaim Weizmann (who later became the first president of Israel) isolated a microorganism that he called BY, which was later named *Clostridium acetobutylicum*. This microorganism is able to ferment starch to acetone, butanol, and ethanol. The first commercial butanol fermentation plant in the United States was built in Terre Haute, Indiana in 1918 by commercial Solvents Corporation using corn as the substrate and *C. acetobutylicum* as the fermenting microorganism. By 1945, the acetone-butanol fermentation was second in importance only to ethanol production by yeast (Dürre, 1998). The ultimate demise of the commercial butanol fermentation process in the United States occurred in the early 1960s due to unfavorable economic conditions brought about by competition with the petrochemical industry. Additionally, butanol fermentation suffered from severe limitations including low product yield, low productivity, and low final product concentrations due to butanol toxicity (Qureshi et al., 1992). However, recent advances in strain development combined with advanced fermentation and product recovery technologies have, at least partially, overcome the above problems (Annous and Blaschek, 1991; Dürre, 1998; Qureshi and Blaschek, 2001a; Ezeji et al., 2003 and 2004). The strain that was used for the commercial production of acetone (butanol was *C. acetobutylicum* P262) (Jones and Woods, 1986). The other species that have been widely studied for the bioconversion of corn to butanol include *C. beijerinckii*, *C. thermosulfurogenes* EM1, *C. saccharolyticum*, and *C. thermosaccharolyticum*.

Recent developments in liquid biofuel technology, uncertainty of petroleum supplies, the finite nature of fossil fuels and environmental concerns have revived research efforts aimed at obtaining liquid fuels from renewable resources. The U.S. Department of Energy has declared that "decreasing U.S dependence on imported oil through the use of biomass-based fuels, power and products is an issue of national security (U.S. Department of Energy 2003)." Butanol is one of the biofuels that has the potential to substitute for gasoline and can be produced from domestically abundant biomass sources including corn. This chapter describes the production of butanol from corn and corn coproducts and the latest developments in butanol production technology including culture development, upstream and downstream processing, and fermentation technology. The formulation of such a chapter is a clear indication that technology to produce butanol from corn is maturing and getting ready for commercialization.

BUTANOL PRODUCTION FROM CORN

AMYLOLYTIC ENZYMES AND SOLVENTOGENIC *CLOSTRIDIA*

The solventogenic *clostridia*, like all *clostridia*, are Gram positive, spore forming, obligate anaerobes. These bacteria can change to a variety of morphologies during fermentation, with motile rod-shaped cells present during the exponential growth phase and dormant oval-shaped endospores formed when the culture encounters adverse conditions. The maintenance of cellular growth (like other heterotrophic bacteria) and butanol production by solventogenic *clostridia* depends on the utilization of nutrients obtained from the surroundings. Corn is principally composed of starch, and starch is made up of amylose and amylopectin. Amylose is composed of a linear polymer of glucose with links exclusively in the α-1, 4 orientation. On the other hand, amylopectin is a highly branched polysaccharide consisting of linear chains of α-1, 4-linked D-glucose residues, joined by α-1, 6-glucosidic bonds. The branch points occur on the average of every 20–25 D-glucose units, so that amylopectin contains 4–5% of α-1, 6-glucosidic linkages (Jensen and Norman, 1984). High-molecular-weight macromolecules like starch from corn are too large to be assimilated by the bacterial cells and therefore need to be hydrolyzed into low-molecular-weight products by specific extracellular depolymerases, which can then be taken into the cells via specific transport systems. Solventogenic *clostridia* have the ability to utilize a wide spectrum of carbohydrates through the secretion of several extracellular amylolytic enzymes.

Several amylolytic enzymes with different modes of action necessary for efficient and complete breakdown of starch to glucose have been identified in the solventogenic *clostridia*. They include α-amylase, β-amylase, glucoamylase, α-glucosidase, and pullulanase, and their mode of action and linkages hydrolyzed in the starch molecule and products formed are summarized in Table 6.1.

Amylases are enzymes that act on starch, glycogen, and derived polysaccharides. They hydrolyze α-1, 4 or α-1, 6 glucosidic bonds between consecutive glucose units. α-Amylase (1,4-α-D-glucanohydrolase; EC 3.2.1.1) catalyzes the hydrolysis of α-1,4 glucosidic bonds in the interior of the substrate molecule (starch, glycogen and various oligosaccharides) and produces a mixture of glucose, maltose, maltotriose, maltotetraose, maltopentose, maltohexaose, and oligosaccharides in a ratio depending on the source of the enzyme (Ezeji, 2001). The β-amylase (1, 4-α-D-glucan maltohydrolase; EC 3.2.1.2) hydrolyzes α-1,4 glucosidic bonds in starch and oligosaccharides producing maltose units from

TABLE 6.1
The Amylolytic Enzymes of the Saccharolytic Solventogenic *Clostridia*

Enzyme	Hydrolyzed Linkages	Mode of Action	Products Formed
α-Amylase	α-1, 4-linkage	Endo-acting (random fashion)	Glucose, linear oligosaccharides and α-limit dextrins
β-Amylase	α-1, 4-linkage	Exo-acting (nonreducing end)	Maltose and β-limit dextrins
Glucoamylase	α-1, 4-linkage and α-1, 6-linkage	Exo-acting (nonreducing end)	Glucose
α-Glucosidase	α-1, 4-linkages	Exo-acting (nonreducing end)	Glucose
Pullulanase	α-1, 6-linkage and pullulan and amylopectin	Endo-acting (random fashion)	Linear oligosaccharides and maltotriose

the nonreducing terminal end of the substrate. Glucoamylase $(1, 4$-α-D-glucan glucohydrolase; EC 3.2.1.3) hydrolyzes both α-1, 4 and α-1, 6 glucosidic linkages from the nonreducing terminal end of the glucose units in the starch molecule. α-Glucosidase (α-D-glucoside glucohydrolase; EC 3.2.1.20) catalyzes, like glucoamylase, the hydrolysis of the terminal nonreducing α-1, 4-linked glucose units in the starch. The preferred substrates for α-glucosidases are maltose, maltotriose, maltotetraose, and short oligosaccharides. Furthermore, pullulanases (α-dextrin 6-glucanohydrolase; EC 3.2.1.41) are enzymes that cleave -1, 6 linkages in pullulan and release maltotriose, although pullulan itself may not be the natural substrate.

Synergistic action between pullulanase and α-amylase enzymes of *C. thermosulfurogenes* has been demonstrated (Spreinat and Antranikian, 1992) and an α-glucosidase of *C. beijerinckii* has been shown to hydrolyze both types of glucosyl linkages (α-1, 4 and α-1, 6) (Albasheri and Mitchell, 1995). In addition, Paquet et al. (1991) purified and characterized novel *C. acetobutylicum* 824 α-amylase, which possesses some glucoamylase activity (2.7%).

BIOCHEMISTRY OF BUTANOL PRODUCTION FROM CORN

Solvent-forming species, including *C. acetobutylicum* and *C. beijerinckii*, are mesophilic, growing best between 30° and 40°C. The pH varies during the fermentation and can drop from an initial value of 6.8–7.0 to about 5.0–4.5 (acidogenesis) and can also rise up to 7.0 later in the fermentation (solventogenesis). It has been suggested that the switch to solvent production is an adaptive response of the cell to the low medium pH resulting from acid production (Bahl et al., 1982).

Solventogenic *clostridia* can be grown on simple media such as ground corn, molasses, whey permeate, or on semidefined and defined media. When semidefined and defined media are used, a wide array of vitamins and minerals are required in addition to a carbohydrate source. *Clostridia* can utilize a wide range of carbohydrates. *C. acetobutylicum* and *C. beijerinckii* can utilize starch, hexoses, pentoses, and cellobiose. Currently, those *clostridia* that are able to utilize cellulose directly produce little or no solvents. Recently, attempts have been made to express cellulase genes in the solventogenic clostridia.

The uptake of carbohydrates in the solventogenic *clostridia* is achieved by a phosphoenolpyruvate (PEP)-dependent phosphotransferase system (PTS). This mechanism involves simultaneous uptake and phosphorylation of substrate that results in the conversion of glucose to glucose-6-phosphate, which is subsequently metabolized to pyruvate via the Embden-Meyerhof-Parnas (EMP) pathway (Mitchell, 2001). Fructose is converted to fructose-1-phosphate and enters the EMP pathway upon conversion to fructose 1,6-bisphosphate. D-xylose is converted to D-xylulose by the xylose isomerase enzyme and the metabolism proceeds by a phosphorylation reaction. The reaction is catalyzed by xylulokinase, which results in the formation of D-xylulose-5-phosphate. The pentose phosphate pathway utilizes enzymes transaldolase and transketolase to convert D-xylulose-5-phosphate to glyceraldehyde-3-phosphate and fructose-6-phosphate (Singh and Mishra, 1995). The glyceraldehyde-3-phosphate and fructose-6-phosphate enter the EMP pathway leading to the formation of pyruvate. The ability of solventogenic *clostridia* to metabolize these sugars is important when corn is considered as the starting material for fermentation, as all of these sugars can be derived from corn or corn coproducts.

Solvent producing *clostridia* metabolize substrates in a biphasic fermentation fashion. During the first phase, acid intermediates (acetic and butyric acids), hydrogen, and a large amount of ATP are produced. In the second phase, butanol, acetone, and ethanol are produced, and hydrogen and ATP production decrease (Jones and Woods, 1986). CO_2 is produced during both phases of growth—two moles are produced from each mole of glucose metabolized to pyruvate—but CO_2 production in the solventogenic phase is higher as an additional mole is produced for every mole of acetone produced. The simplified overall fermentation pathway is given in Figure 6.1.

During the acidogenic phase, cells typically grow exponentially due to the high amount of ATP (3.25 mol/mol of glucose) being produced (Jones and Woods, 1986). The enzymes phosphate acetyltransferase and acetate kinase convert acetyl-CoA to acetate and, analogously, phosphate butyltransferase and butyrate kinase convert butyryl-CoA to butyrate during this phase of growth. The pH of the fermentation broth decreases as butyric and acetic acids accumulate. The acetic and butyric acids produced during the fermentation may be freely permeable to the cell membrane and these acids equilibrate the internal (bacterial) and fermentation broth pH. Both reduction of pH and accumulation of acetate and butyrate have been associated with triggering solventogenesis (Jones and Woods, 1986).

The solventogenic phase is typically associated with stationary phase. ATP production is reduced to 2 mol/mol of glucose during this phase. The fermentation intermediates (acetic and butyric acids) are reassimilated and converted into acetone and butanol. It has been suggested that butyric and acetic acids are reassimilated by the action of the enzyme acetoacetyl-CoA:acetate/butyrate:CoA transferase (Andersch et al., 1983). This enzyme catalyzes the reaction that transfers CoA from acetoacetyl-CoA to either acetate or butyrate. Acetate is converted to acetyl-CoA, which can be converted to acetone, butanol, or ethanol. Butyrate is converted to butyryl-CoA, which can only be used to produce butanol. This is because there is no metabolic pathway to regenerate acetyl-CoA from butyryl-CoA. When CoA is removed from acetoacetyl-CoA, acetoacetate is produced, which can be transformed directly into acetone and CO_2 by acetoacetate decarboxylase.

The central core of both the acidogenic and solventogenic pathways is the series of reactions that produces butyryl-CoA from acetyl-CoA. Thiolase condenses two molecules of acetyl-CoA into one molecule of acetoacetyl-CoA. Acetoacetyl-CoA is reduced to 3-hydroxybutyryl-CoA by hydroxybutyryl-CoA dehydrogenase. From this, crotonyl-CoA is formed by dehydration, catalyzed by

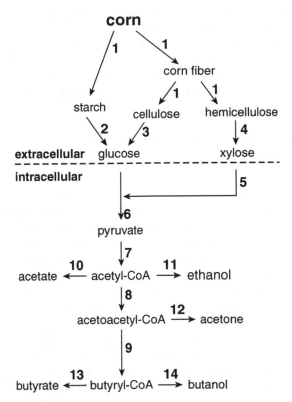

FIGURE 6.1 Simplified metabolism of polysaccharides by solventogenic *clostridia*. Symbols: 1, Pretreatment of corn; 2, Starch hydrolysis; 3, Cellulose hydrolysis; 4, Hemicellulose hydrolysis; 5, Xylose uptake and subsequent breakdown breakdown via the transketolase-transaldolase sequence; 6, Glucose uptake by the phosphotransferase system (PTS) and conversion to pyruvate by the Embden-Meyerhof-Parnas (EMP) pathway; 7, pyruvate-ferrodoxin oxidoreductase; 8, thiolase; 9, 3-hydroxybutyryl-CoA dehydrogenase, crotonase and butyryl-CoA dehydrogenase; 10, phosphate acetyltransferase and acetate kinase; 11, acetaldehyde dehydrogenase and ethanol dehydrogenase; 12, acetoacetyl-CoA:acetate/butyrate:CoA transferase and acetoacetate decarboxylase; 13, phosphate butyltransferase and butyrate kinase; 14, butyraldehyde dehydrogenase and butanol dehydrogenase.

crotonase. The carbon-carbon double bond in crotonyl-CoA is reduced with NADH to produce butyryl-CoA. This last step is catalyzed by butyryl-CoA dehydrogenase (Bennett and Rudolph, 1995).

   The accumulation of both acids (butyrate and acetate) and solvents (acetone, butanol, and ethanol) in the fermentation broth is toxic to the microorganism and eventually causes cell death. The shift to solventogenesis is effective in extending the fermentation, but the butanol produced eventually reaches toxic levels. The presence of butanol in the membrane increases membrane fluidity and destabilizes the membrane and membrane-associated processes (Jones and Woods, 1986). The maximum amount of solvents (total acetone, butanol, and ethanol) that the cell can tolerate is 20 gL⁻¹ (Maddox, 1989). This limits the amount of glucose that can be fermented in batch culture to 60 gL⁻¹ because using a higher concentration of glucose would result in incomplete substrate utilization due to butanol toxicity. Many studies today are focused on overcoming the butanol toxicity issue, whether by developing a more butanol tolerant microorganism or by selectively removing butanol from the fermentation broth.

BUTANOL PRODUCTION FROM CORN COPRODUCTS

Corn Fiber

Corn fiber is a coproduct of the corn wet-milling industry. It is a mixture of corn kernel hulls and residual starch not extracted during the wet-milling process. Corn fiber is composed of approximately 40% hemicellulose, 12% cellulose, 25% starch, 10% protein, 3% oil, and 10% other substances such as ash and lignin (Singh et al., 2003). Corn fiber represents a renewable resource that is available in significant quantities from the corn dry- and wet-milling industries. Approximately 6.3×10^6 dry tons of corn fiber is produced annually in the United States. Typically 4.5 lb of corn fiber is obtained from a bushel (56 lb) of corn, which can be converted to about 3.0 lb of fermentable sugars. The major fermentable sugars from hydrolysis of lignocellulosic biomass, such as softwood, hardwood and grasses, rice and wheat straw, sugarcane bagasse, corn stover and corn fiber, are D-glucose and D-xylose (except that softwood also contains substantial amounts of mannose) (Sedlak and Ho, 2004). Industrial *Saccharomyces* yeast strains used for fermenting sugars to ethanol lack the ability to utilize xylose, one of the major end products of hemicellulose hydrolysis. This is a major obstacle for the utilization of corn fiber or other forms of lignocellulosic-based biomass.

Economically, it is important that both xylose and glucose present in corn fiber be fermented to butanol in order for this renewable biomass to be used as feedstock for butanol production. Solventogenic *clostridia* have an added advantage over many other cultures as they can utilize both hexose and pentose sugars (Singh and Mishra, 1995) released from lignocellulosic biomass upon hydrolysis to produce butanol. Fond and Engasser (1986), during their evaluation of the fermentation of lignocellulosic hydrolysates to butanol by *C. acetobutylicum* ATCC 824, demonstrated that the culture utilized both xylose and glucose, although xylose was utilized more slowly than glucose and also supported lower butanol production. However, *C. beijerinckii* BA101 has been shown to utilize xylose and can effectively coferment xylose and glucose to produce butanol (Ebener et al., 2003). Parekh et al. (1988) produced acetone-butanol from hydrolysates of pine, aspen, and corn stover using *C. acetobutylicum* P262. Similarly Marchal et al. (1984) used wheat straw hydrolysate and *C. acetobutylicum,* while Soni et al. (1982) used bagasse and rice straw hydrolysates and *C. saccharoper-butylacetonicum* to convert these agricultural wastes into butanol.

An important limitation of corn fiber utilization comes from the pretreatment and hydrolysis of corn fiber to glucose and xylose. Saccharification of corn fiber can readily be achieved by treatment with dilute H_2SO_4. However, this acid-catalyzed reaction leads to the degradation of glucose to hydroxy methyl furfural (HMF) and xylose to furfural at the temperatures of hydrolysis, resulting in inhibition of fermentation by these degradation products. Other degradation products include syringaldehyde, acetic, ferulic, and glucuronic acids. The formation of these degradation products lowers the yield of fermentable sugars obtained from the corn fiber and the degradation products are inhibitory to yeast and bacterial fermentations. *C. beijerinckii* BA101 is able to completely utilize enzyme-hydrolyzed corn fiber to produce acetone-butanol, but performed poorly in the bioconversion of acid-hydrolyzed corn fiber to acetone-butanol due to the presence of inhibitory compounds generated during hydrolysis (Ebener et al., 2003). Therefore, the development of strains that can tolerate the inhibitory compounds generated during acid pretreatment and hydrolysis of corn fiber remains a priority.

Corn Steep Water

Corn steep water (CSW) is a by-product of the corn wet-milling industry and contains large amounts of substances derived from the fermentative conversion of carbohydrates, proteins, and lipids during corn steeping. Currently, CSW is evaporated to 50% solids and marketed primarily as an economical livestock feed supplement in the cattle industry. CSW is a rich complement of important nutrients such as nitrogen, amino acids, vitamins, and minerals and was proposed to be a good substitute for yeast extract (Hull et al., 1996). This finding is important as it impacts the economics of butanol production from corn.

An economic analysis performed by Qureshi and Blaschek (2000a), demonstrated that the fermentation substrate was one of the most important factors that influenced the price of butanol. Development of a cost-effective biomass-to-butanol process can only be commercially viable if cheaper commercial substrate such as liquefied corn starch and CSW can be used in combination with toxic product removal by gas stripping (Ezeji et al., 2005). It is interesting to note that *C. beijerinckii* BA101 when grown on liquefied corn starch-CSW medium produced levels of acetone-butanol equal to or higher than the levels produced when grown on glucose-based yeast extract medium (Ezeji et al., 2005). The fermentation time for liquefied corn starch- and saccharified liquefied corn starch-CSW media were 120 and 78 h, respectively, while the fermentation time for glucose-based yeast extract medium was 68 h. The presence of sodium metabisulfite ($Na_2S_2O_5$; a preservative) in the liquefied starch and CSW was found to result in inhibition of *C. beijerinckii* BA101 and also may have affected the secretion of amylolytic enzymes by the culture, which is necessary for efficient hydrolysis and utilization of starch and oligosaccharides. However, it appears that the use of CSW has a great potential for the bioconversion of corn to acetone-butanol. The presence of $Na_2S_2O_5$ in the CSW may be a major problem in a long-term fermentation by *C. beijerinckii* BA101. During a long-term fermentation using CSW and *C. beijerinckii* BA101, removal of $Na_2S_2O_5$ from CSW by oxidation is recommended (Ezeji et al., 2005).

BUTANOL PRODUCTION PROCESSES

Batch Process

Batch fermentation is the most commonly studied process for butanol production. In the batch process the substrate (feed) and nutrients are charged into the reactor that can be used by the culture. In a batch process, a usual substrate concentration of 60–80 gL^{-1} is used as higher concentration results in residual substrate being in the reactor. The reaction mixture is then autoclaved at 121°C for 15 minutes followed by cooling to 35–37°C and inoculation with the seed culture. During cooling, nitrogen, or carbon dioxide is swept across the surface to keep the medium anaerobic. After inoculation, the medium is sparged with these gases to mix the inoculum. Details of seed development and inoculation have been published elsewhere (Formanek et al., 1997; Qureshi and Blaschek, 1999a). Depending on the size of the final fermentor, the seed may have to be transferred several times before it is ready for the production fermentor.

Various substrates can be used to produce butanol including corn, molasses, whey permeate, or glucose derived from corn (Qureshi and Blaschek, 2005). However, some substrates may require processing prior to fermentation, known as "upstream processing," such as dilution, concentration, centrifugation, filtration, hydrolysis, etc. The usual batch fermentation time lasts from 48 to 72 h after which butanol is recovered, usually by distillation. During this fermentation period, ABE up to 33 gL^{-1} is produced using hyperbutanol producing *C. beijerinckii* BA101 (Chen and Blaschek, 1999; Formanek et al., 1997). This culture results in a solvent yield of 0.40–0.42 (Formanek et al., 1997). The ABE concentration in the fermentation broth is limited due to butanol inhibition to the cell. At a butanol concentration of approximately 20 gL^{-1}, strong cell growth inhibition occurs that kills the cells and stops the fermentation. Butanol production is a biphasic fermentation where acetic and butyric acids are produced during acidogenic phase followed by their conversion into acetone and butanol (solventogenic phase). During the acidogenic phase, the pH drops due to acid production and subsequently rises during solvent production. At the end of fermentation, cell mass and other suspended solids (if any) are removed by centrifugation and sold as cattle feed. Figure 6.2 shows fermentation profile of butanol production in a typical batch fermentation process from cornstarch using *C. beijerinckii* BA101.

Butanol can be produced both by using corn coproduct from i) corn dry-grind and ii) wet-milling processes. During the dry-grind process corn fiber and germ are not removed prior to fermentation. At the end of fermentation (after starch utilization during fermentation), corn fiber and other insoluble solids are removed by centrifugation, dried, and sold as cattle feed. The dried solids are known as "Distillers Dry Grain Solids" or DDGS. On the contrary, during the wet-milling

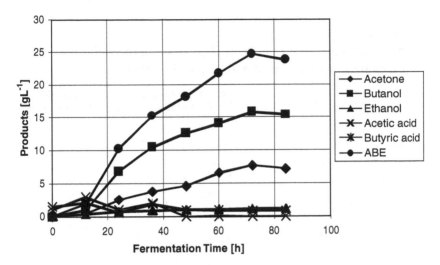

FIGURE 6.2 Fermentation profile of ABE production from 60 gL⁻¹ cornstarch in a batch reactor using *C. beijerinckii* BA101.

process, corn fiber and germ are removed prior to fermentation. In this process, cornstarch can be converted to any of the three products (liquefied cornstarch, glucose syrup, or glucose) each of which is fermentable by *C. beijerinckii* to produce butanol. It should be noted that often corn refineries add sodium metabisulfite during the wet-milling operation as a corn kernel softening agent and preservative to the liquefied cornstarch. The presence of sodium metabisulfite may interfere with the direct fermentation of the liquefied cornstarch. However, glucose syrup or glucose does not contain any such fermentation inhibitors. The unit operations that are applicable to the corn dry-grind and wet-milling fermentation of butanol are given in Table 6.2.

During the 1940s and 1950s, production of butanol on an industrial scale (Terre Haute, IN, and Peoria, IL) was carried out using large fermenters ranging in capacity from 200,000 to 800,000 L. The industrial process used 8–10% corn mash, which was cooked for 90 min at 130–133°C. Corn contains approximately 70% (dry weight basis) starch. The use of molasses offers many advantages over using corn, including the presence of essential vitamins and micronutrients (Paturau, 1989). In industrial processes, beet and invert and blackstrap molasses were diluted to give a fermentation sugar concentration of 50 to 75 gL⁻¹, most commonly 60 gL⁻¹. The molasses solution was sterilized at 107 to 120°C for 15 to 60 min followed by adding organic and inorganic nitrogen, phosphorus, and buffering chemicals. The yield of solvent using *C. acetobutylicum* was usually low at 0.29–0.33. Distillation has been the method of choice to recover butanol; however, during the last two decades a number of alternative techniques have been investigated for the economical recovery of butanol, which will be discussed in the recovery section.

TABLE 6.2
Unit Operations That Can Be Applied to Butanol Production Employing Corn Dry-Grind and Wet-Milling Processes

Unit Operations	Butanol Production by Dry-Grind Process	Butanol Production by Wet-Milling Process
Batch fermentation	x	x
Batch process with concentrated feed	–	x
Fed-batch fermentation	–	x
Continuous fermentation	Difficult due to starch viscosity	x
Immobilized cell fermentation	–	x
Cell recycle	–	x
Recovery by gas stripping	Possible	x
Pervaporation	Possible; solids separation prior to recovery essential	x
Liquid-liquid extraction	Same as above	x
Perstraction	Same as above	x

Note: Numbers before unit operations are section numbers in this chapter.
x - applicable
– Not possible

Batch Process with Concentrated Sugar Solutions

Due to the toxic nature of butanol, the initial substrate concentration is limited to <80 gL^{-1} (usually 60 gL^{-1}). A substrate concentration in excess of this results in a high residual substrate, thus resulting in inefficient sugar utilization and increased BOD (biological oxygen demand) load for wastewater treatment. However, recent developments in downstream processing (recovery) of ABE have made it possible to use concentrated sugar solutions for this fermentation. During the fermentation, the toxic products are removed simultaneously, thus relieving inhibition that results in the utilization of more substrate. The details of the recovery techniques are given in the recovery section (3.1). Employing butanol removal techniques, sugar solutions containing 161 gL^{-1} glucose (*C. beijerinckii*; Ezeji et al., 2003) and 227 gL^{-1} lactose (*C. acetobutylicum*; Qureshi and Maddox, 2005) have been successfully used. Use of concentrated glucose and lactose solutions has resulted in the production of 76 and 137 gL^{-1} ABE, respectively. In such fermentations, fewer acids are produced, thus improving the ABE yield. In another process, 200 gL^{-1} lactose was successfully fermented in a batch reactor of *C. acetobutylicum* when integrated with product recovery by gas stripping (Maddox et al., 1995). This system resulted in the production of 70 gL^{-1} ABE with a productivity of 0.32 gL^{-1}h^{-1} as compared to 0.07 gL^{-1}h^{-1} in the control batch reactor. Studies reported in this section demonstrated that a fermentation medium containing over three times the sugar concentration as compared to a batch reactor can be successfully fermented when integrated with product removal techniques.

Fed-Batch Fermentation

Fed-batch fermentation is a technique that is applied to processes in which a high substrate concentration is toxic to the culture. In such a case, the reactor is initiated in a batch mode with a low substrate concentration (usually 60–100 gL^{-1}) and low fermentation medium volume, usually less than half the volume of the fermentor. The reactor is inoculated with the culture and the fermentation proceeds. As the substrate is utilized by the culture, it is replaced by adding a concentrated substrate solution at a slow rate, thereby keeping the substrate concentration in the fermentor below the toxic level to the culture (Ezeji et al., 2004). When using this approach, the culture volume increases over time unless culture fluid is removed. The culture is harvested when the liquid volume is approximately 75% of the volume of reactor. Since butanol is toxic to *C. aceto-butylicum* and *C. beijerinckii* cells, the fed-batch fermentation technique cannot be applied in this case unless one of the novel simultaneous fermentation and product recovery techniques is applied. In a number of studies (Ezeji et al., 2004; Qureshi et al., 1992), this technique has been applied successfully to the ABE fermentation. In fed-batch fermentation, Ezeji et al. (2004) were able to utilize 500 g glucose in 1 L culture volume (500 gL^{-1}) as compared to 60 gL^{-1} in a control batch process.

Qureshi et al. (1992) used a concentrated substrate of whey permeate (350 gL^{-1}) to produce butanol in a fed-batch reactor of *C. acetobutylicum*. In this process, three different techniques of butanol separation were compared including perstraction, gas stripping, and pervaporation. In the three processes, 57.8, 69.1 and 42.0 gL^{-1} ABE were produced, respectively. ABE yield of 0.37, 0.38, and 0.34 were obtained, respectively. Overall, application of these three techniques suggested that fed-batch fermentation technique can be applied to the ABE fermentation provided ABE is removed from the culture broth simultaneously. In another study, Qureshi et al. (2001) produced ABE in a fed-batch fermentation of *C. acetobutylicum* and removed these solvents using a silicone-silicalite synthesized pervaporation membrane. The reactor was fed with 700 gL^{-1} glucose solution. In this system, 155 gL^{-1} ABE was produced with an average yield of 0.31–0.35 and productivities ranging from 0.13 to 0.26 $gL^{-1}h^{-1}$. Using the fed-batch technique, fermentation of more sugars (2–3 times) was possible when novel product removal techniques were applied to the process.

Continuous Fermentation

The continuous culture technique is often used to improve reactor productivity and to study the physiology of the culture in steady state. A number of studies exist for continuous fermentation of butanol, and they all give some insight into butanol fermentation and the behavior of the culture under these conditions. Because of the production of fluctuating levels of solvents and the complexity of butanol fermentation, the use of a single-stage continuous reactor does not seem to be practical at the industrial scale. In continuous culture, a serious problem exists in that solvent production may not be stable for long time periods and ultimately declines over time, with a concomitant increase in acid production. In a single-stage continuous system, high reactor productivity may be obtained, however, at the expense of low product concentration compared to that achieved in a batch process. In a single-stage continuous reactor using *C. acetobutylicum*, Leung and Wang (1981) produced 15.9 gL^{-1} total solvents (ABE) at a dilution rate of 0.1 h^{-1} resulting in a productivity of 1.6 $gL^{-1}h^{-1}$. The productivity was improved further to 2.55 $gL^{-1}h^{-1}$ by increasing the dilution rate to 0.22 h^{-1}. It should be noted that the product concentration decreased to 12.0 gL^{-1}. In a related continuous fermentation process using a hyperbutanol producing strain of *C. beijerinckii* BA101, Formanek et al. (1997) was able to produce 15.6 gL^{-1} ABE at a dilution rate of 0.05 h^{-1} resulting in a productivity of 0.78 $gL^{-1}h^{-1}$. However, solvent concentration decreased to 8.7 gL^{-1} as dilution rate was increased to 0.2 h^{-1}. This resulted in an increase in productivity to 1.74 $gL^{-1}h^{-1}$.

As a means of increasing product concentration in the effluent and reducing fluctuations in butanol concentration, two or more multistage continuous fermentation systems have been investigated (Bahl et al., 1982; Yarovenko, 1964). Often, this is done by allowing cell growth, acid production, and ABE production to occur in separate bioreactors. In a two-stage system, Bahl et al. (1982) reported a solvent concentration of 18.2 gL^{-1} using *C. acetobutylicum* DSM 1731, which is comparable to the solvent concentration in a batch reactor. This type of multistage bioreactor system (7–11 fermenters in series) was successfully tested at the pilot scale and full plant scale level in the Soviet Union (now Russia) (Yarovenko, 1964). However, 7–11 fermenters in series add to the complexity of the system for a relatively low-value product such as butanol. It is viewed that such a multistage system would not be economical.

Immobilized and Cell Recycle Reactors

Increased reactor productivity results in the reduction of process vessel size and capital cost thus improving process economics. In a butanol batch process, reactor productivity is limited to less than 0.50 $gL^{-1}h^{-1}$ due to a number of reasons including low cell concentration, down time and product inhibition (Maddox, 1989). Increasing cell concentration in the reactor is one of the methods to improve reactor productivity. Cell concentration can be increased by one of two techniques namely, "immobilization" and "cell recycle." In a batch reactor a cell concentration of <4 gL^{-1} is normally achieved. In an attempt to improve the reactor productivity, Ennis et al. (1986a) were among the early investigators to use the cell immobilization technique for the butanol fermentation. These authors used cell entrapment technique and continuous fermentation with limited success in productivity improvement. The same group investigated another technique involving cell immobilization by adsorption onto bonechar and improved reactor productivity to approximately 4.5 $gL^{-1}h^{-1}$ (Qureshi and Maddox, 1987) followed by further improvement to 6.5 $gL^{-1}h^{-1}$ (Qureshi and Maddox, 1988). The culture that was used in these studies was *C. acetobutylicum* P262. In an attempt to explore clay bricks as an adsorption support for cells of *C. beijerinckii*, Qureshi et al. (2000) were able to improve reactor productivity to 15.8 $gL^{-1}h^{-1}$. In another approach, Huang et al. (2004) immobilized cells of *C. acetobutylicum* in a fibrous support, which was used in a continuous reactor to produce ABE. In this reactor a productivity of 4.6 $gL^{-1}h^{-1}$ was obtained.

Cell recycle technique is another approach to increase cell concentration in the reactor and improve reactor productivity (Cheryan, 1986). Using this approach, reactor productivities up to 6.5 $gL^{-1}h^{-1}$ (as compared to <0.5 $gL^{-1}h^{-1}$ in batch fermentation) have been achieved in the butanol fermentation (Afschar et al., 1985; Pierrot et al., 1986). In a similar approach, Mulchandani and Volesky (1994) used a single-stage spin filter perfusion bioreactor in which a maximum productivity of 1.14 g L^{-1} h^{-1} was obtained; however, the ABE concentration fluctuated over time.

ENHANCEMENT OF SUBSTRATE UTILIZATION AND BUTANOL PRODUCTIVITY

NOVEL DOWNSTREAM PROCESSING

Gas Stripping

Gas stripping is a simple technique that can be applied for recovering butanol (ABE) from the fermentation broth (Maddox, 1989; Qureshi and Blaschek, 2001b). Oxygen-free nitrogen or fermentation gases (CO_2 and H_2) are bubbled through the fermentation broth followed by cooling the gas (or gases) in a condenser. As the gas is bubbled through the fermentor, it captures ABE, which is condensed in the condenser followed by collection in a receiver. Once the solvents are condensed, the gas is recycled back to the fermentor to capture more ABE. This process continues until all the sugar in the fermentor is utilized by the culture. In some cases, a separate stripper can be used to strip off solvents followed by recycling the stripper effluent that is low in ABE. Figure 6.3 shows a typical schematic diagram of solvent removal by gas stripping. Gas stripping has been successfully applied to remove solvents from batch (Ennis et al., 1986b; Maddox et al., 1995; Ezeji et al., 2003), fed-batch (Qureshi et al., 1992, Ezeji et al., 2004), fluidized bed (Qureshi and Maddox, 1991a) and continuous reactors (Groot et al., 1989; Ezeji et al., 2002). In addition to removal of solvents, a concentrated sugar solution was fed to the reactors to reduce the volume of process streams and economize the butanol production process. The reader is referred to the *Batch process with concentrated sugar solutions* and *fed-batch fermentation* sections of this chapter where concentrated sugar solutions were successfully fermented in combination with product recovery by gas stripping. In these processes, the reactor productivities and product yield were also improved (Maddox

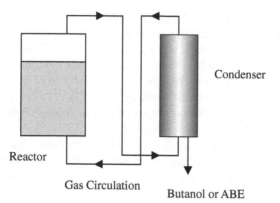

Condenser

Reactor

Gas Circulation   Butanol or ABE

FIGURE 6.3 A schematic diagram of ABE recovery from fermentation broth by gas stripping.

et al., 1995; Qureshi and Maddox, 1991a; Ezeji et al., 2002). Additional advantages of gas stripping included achieving a high product concentration in the condensed stream and the requirement for no membrane or chemicals for the recovery process.

Pervaporation

Pervaporation is a technique that allows selective removal of volatiles from model solution/fermentation broths using a membrane. The volatile or organic component diffuses through the membrane as a vapor followed by recovery by condensation. In this process, a phase change occurs from liquid to vapor. Since it is a selective removal process, the desired component requires a heat of vaporization at the feed temperature. The mechanism by which a volatile/organic component is removed by pervaporation is called solution-diffusion. In pervaporation, the effectiveness of separation of a volatile is measured by two parameters called selectivity (a measure of selective removal of volatile) and flux (the rate at which an organic/volatile passes through the membrane per m² membrane area). A schematic diagram of the pervaporation process is shown in Figure 6.4. The details of pervaporation have been described in the literature (Maddox, 1989; Groot et al., 1992; Qureshi and Blaschek, 1999b).

Application of pervaporation to batch butanol fermentation has been described by Groot et al. (1984), Larrayoz and Puigjaner (1987), Qureshi and Blaschek (1999a), and Fadeev et al. (2001). Pervaporation has also been used for the removal of butanol from the fermentation broth in fed-batch reactors (Qureshi and Blaschek, 2000b; Qureshi and Blaschek, 2001a). In the fed-batch reactors concentrated sugar solutions have been used to reduce the process stream volume. It is interesting to note that acids did not diffuse through the membranes used by the above authors. Qureshi et al. (1992) used a polypropylene membrane through which diffusion of acids occurred, however, at high acid concentration in the fermentation broth.

Broth Recycle

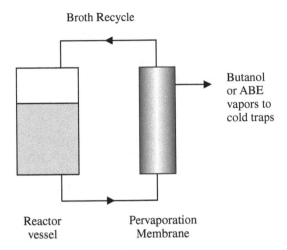

Butanol
or ABE
vapors to
cold traps

Reactor   Pervaporation
vessel   Membrane

FIGURE 6.4 A schematic diagram of ABE recovery from fermentation broth by pervaporation.

In an attempt to improve membrane selectivity, Matsumura et al. (1988) applied a combination of liquid-liquid extraction and pervaporation to recover butanol. The extraction solvent used for this process was oleyl alcohol, which formed a liquid layer (also known as a thin membrane) on a microporous 25 μm thick polypropylene flat sheet. The oleyl alcohol also got impregnated into the sheet pores. In this combination of liquid-solid membrane, oleyl alcohol dissolved butanol relatively quickly followed by diffusion through the polypropylene membrane. The advantage of combining the liquid and solid membrane was that a high butanol selectivity (180) was achieved in comparison to a low selectivity (10–15) when using polypropylene film alone. It was estimated that if this solid-liquid pervaporation membrane were used for butanol separation, the energy requirement would be only 10% of that required in a conventional distillation. Unfortunately, the membrane was not stable as the oleyl alcohol that formed a thin film and was impregnated into the polypropylene film pores diffused out of the membrane.

In order to develop a stable and highly selective membrane, Qureshi et al. (1999) employed two techniques known as adsorption and pervaporation. It has been reported that adsorption of butanol onto silicalite and molecular sieves is a quick and selective process (Ennis et al., 1987). Qureshi et al. (1999) synthesized a membrane in which silicalite, an adsorbent, was included into a silicone membrane. By combining these, butanol selectivity was improved from 40 (silicone membrane) to 209. The membrane developed was called a silicalite-silicone membrane. The membrane was found to be stable with a working life of three years. This membrane was used with both butanol model solutions and fermentation broths (Qureshi et al., 1999; 2001). A comparison of various membranes suggested that this membrane may be superior to other pervaporation membranes used for butanol separation (Qureshi and Blaschek, 1999b). Some of the details of the use of this membrane have been given in the section *fed-batch fermentation* (Qureshi et al., 2001) where 155 gL^{-1} ABE was produced in the integrated fermentation and product recovery process as compared to <20 gL^{-1} in a batch process. This membrane was so efficient that a butanol concentration up to 700 gL^{-1} was achieved in the permeate.

Liquid-Liquid Extraction

Liquid-liquid extraction has been considered an important technique for the recovery of ABE from fermentation broths. Usually, a water-insoluble organic extractant is mixed with the fermentation broth. Butanol is more soluble in the organic (extractant) phase than in the aqueous (fermentation broth) phase. Therefore, butanol selectively concentrates in the organic phase. Since the extractant and fermentation broth are immiscible, the extractant can easily be separated from the fermentation broth after butanol extraction. Liquid-liquid extraction is able to remove fermentation products without removing substrates, water, or nutrients.

In order to improve substrate utilization and productivity, liquid-liquid extraction must be integrated with butanol fermentation such that simultaneous fermentation and butanol removal from the fermentation broth is achieved. The choice of extractant is critical because an extractant with low partition coefficient will not be efficient in the recovery of butanol and a toxic extractant will inhibit or kill the bacterial cells. Unfortunately, most extractants with high partition coefficient are toxic to the *clostridia*. The extractant of choice among researchers has been oleyl alcohol because it is nontoxic and a good extractant as well (Evans and Wang, 1988, Groot et. al., 1990).

Perstraction

Perstraction is a butanol recovery technique similar to liquid-liquid extraction that seeks to solve some of the problems inherent in liquid-liquid extraction. In a perstraction separation, the fermentation broth and the extractant are separated by a membrane (Qureshi et. al., 1992). The membrane contactor provides a surface area where the two immiscible phases can exchange the butanol. Since there is no direct contact between the two phases, extractant toxicity, phase dispersion, emulsion and rag layer formation are drastically reduced or eliminated. In such a system, butanol should diffuse preferentially across the membrane, while other components such as medium compositions and fermentation intermediates (acetic and butyric acids) should be retained in the aqueous phase. The total mass transport of butanol from the fermentation broth to the organic side depends on the rate of diffusion of butanol across the membrane. The net movement is measured as membrane flux or rate of movement, $J = dQ_b/dt$, where J = net flux and dQ_b/dt = diffusion rate (influx + efflux) of butanol. The membrane does, however, present a physical barrier that can limit the rate of solvent extraction.

ECONOMIC SCENARIOS

ECONOMICS OF BUTANOL PRODUCTION

In recent years a number of economic studies have been performed on the production of butanol from corn (Marlatt and Datta, 1986; Qureshi and Blaschek, 2000a; 2001a), whey permeate (Qureshi and Maddox, 1991b; 1992), and molasses (Qureshi and Maddox, 1991b; 1992). In these studies, it has been determined that distillative recovery of butanol from fermentation broth is not economical as compared to butanol derived from petrochemicals (current route). It has also been identified that new developments in process technology for butanol production from renewable substrates allows for a significant reduction in the price of butanol. The price of butanol derived from corn also depends upon the coproducts credit, which is significant. Currently, it is anticipated that the petrochemical industries would reduce the price of butanol in an attempt to prevent the fermentative production of butanol from being successful. At present, the petrochemical industries have a monopoly with respect to the butanol market.

To bring fermentatively derived butanol closer to commercialization and compete with petrochemically derived butanol, it is suggested that research be focused on the development of superior cultures (as compared to the existing strains: *C. beijerinckii* BA101 and *C. acetobutylicum* PJC4BK). These cultures produce total ABE on the order of 25-33 gL^{-1} (Formanek et al., 1997; Chen and Blaschek, 1999; Harris et al., 2000). Further improvements in ABE yield, which is 0.40–0.42 when using *C. beijerinckii* BA101, should also be examined. Other cultures have been reported to result in a product yield of approximately 0.30. Material balance suggests that approximately 53% of carbon is lost as CO_2, indicating that only 47% of the substrate is directed for the product conversion.

Another problem with butanol fermentation is the inability of these cultures to use sugars derived from economically available substrates such as corn fiber hydrolysate (Ebener et al., 2003). As with corn fiber hydrolysate, it is anticipated that sugars derived from hydrolysed corn stalks, wheat straw, and rice husk would not be utilized without pretreatment of these substrates, which would further add to the processing cost. In order to meet these challenges, new strains capable of utilizing agricultural biomass derived sugars should be developed. Alternately, economic methods capable of removing inhibitors from the hydrolysates should be developed. Simultaneous saccharification and fermentation is another approach that should be investigated for this process.

CONCLUSION

The production of butanol via the fermentation route is a relatively complicated process because the solventogenic *clostridia* are obligate anaerobes and the fermentation product (butanol) is toxic to the producing cultures. The possibility of incorporating in-line product recovery processes such as liquid-liquid extraction, perstraction, pervaporation, and gas stripping has generated a lot of interest.

Simultaneous butanol fermentation and recovery has dramatically improved the productivity of butanol production from corn. By employing in-line recovery systems during butanol fermentation, substrate inhibition and butanol toxicity to the culture are drastically reduced. Given that butanol is an excellent potential fuel and the United States is rich in biomass, butanol production from corn has a bright future. As it is seen at this stage, the technology of butanol production from corn (and other substrates) is ready for commercialization; however, this also depends upon the fluctuations in crude oil prices.

REFERENCES

Afschar, A.S., Biebl, H., Schaller, K. and Schugerl, K., Production of acetone and butanol by *Clostridium acetobutylicum* in continuous culture with cell recycle, *Appl. Microbiol. Biotechnol.*, 22, 394–398, 1985.

Albasheri, K.A. and Mitchell, W.J., Identification of two α-glucosidase activities in *Clostridium acetobutylicum* NCIB 8052, *J. Appl. Bacteriol.*, 78, 149–156, 1995.

Andersch, W., Bahl, H. and Gottschalk, G., Level of enzymes involved in acetate, butyrate, acetone and butanol formation by *Clostridium acetobutylicum*, *Eur. J. Appl. Microbiol. Biotechnol.*, 18, 327–332, 1983.

Annous, B.A. and Blaschek, HP., Isolation and characterization of *Clostridium acetobutylicum* mutants with enhanced amylolytic activity, *Appl. Environ. Microbiol.*, 57, 2544–2548, 1991.

Bahl, H., Andersch, W. and Gottschalk, G., Continuous production of acetone and butanol by *Clostridium acetobutylicum* in a two-stage phosphate limited chemostat, *Eur. J. Appl. Microbiol. Biotechnol.*, 15, 201–205, 1982.

Bennett, G.N. and Rudolph, F.B., The central metabolic pathway from acetyl-CoA to butyryl-CoA in *Clostridium acetobutylicum*, *FEMS Microbiol. Rev.*, 17, 241–249, 1995.

Chen, C.K. and Blaschek, H.P., Acetate enhances solvent production and prevents degeneration in *Clostridium beijerinckii* BA101, *Appl. Microbiol. Biotechnol.*, 52, 170–173, 1999.

Cheryan, M., *Ultrafiltration Handbook*, Technomic Press, Lancaster, PA, 1986.

Dürre, P., New insights and novel developments in clostridial acetone/butanol/isopropanol fermentation, *Appl. Microbiol. Biotechnol.*, 49, 639–648, 1998.

Ebener, J., Qureshi, N., Ezeji, T.C. and Blaschek, H.P. Corn fiber hydrolysis and fermentation to butanol using *Clostridium beijerinckii* BA101, In *Proceedings of 25th Biotechnology Symposium on Fuels and Chemicals*, Breckenridge, CO, May 4–7, 2003, p. 65.

Ennis, B.M., Maddox, I.S. and Schoutens, G.H. Immobilized *Clostridium acetobutylicum* for continuous butanol production from whey permeate, *New Zealand J. Dairy Sci. Technol.*, 21, 99–109, 1986a.

Ennis, B.M., Marshall, C.T., Maddox, I.S. and Paterson, A.H.J., Continuous product recovery by *in-situ* gas stripping/condensation during solvent production from whey permeate using *Clostridium acetobutylicum*, *Biotechnol. Lett.*, 8, 725–730, 1986b.

Ennis, B.M., Qureshi, N. and Maddox, I.S., Inline toxic product removal during solvent production by continuous fermentation using immobilized *Clostridium acetobutylicum*, *Enzym. Microbial Technol.*, 9, 672–675, 1987.

Evans, P.J. and Wang, H.Y., Enhancement of butanol formation by *Clostridium acetobutylicum* in the presence of decanol-oleyl alcohol mixed extractants, *Appl. Env. Microbiol.*, 54, 1662–1667, 1988.

Ezeji, T.C., Production, purification and characterization of thermostable amylolytic enzymes from the newly isolated *Bacillus thermodenitrificans* HRO10. Ph.D. Dissertation. Der Andere Verlag, Kollegienwall, Osnabruek, Germany, 2001, pp. 1–125.

Ezeji, T.C., Qureshi, N. and Blaschek, H.P., Continuous production of butanol using *Clostridium beijerinckii* BA101 and recovery by gas stripping, *Clostridium* 2002: The Seventh International Conference and Workshop on Regulation of Metabolism, Genetics and Development of the Solvent and Acid Forming *Clostridia*, Rostock, Germany, Sept. 19–21, 2002.

Ezeji, T.C., Qureshi, N. and Blaschek, H.P., Production of butanol by *Clostridium beijerinckii* BA101 and in-situ recovery by gas stripping, *World J. Microbiol. Biotechnol.*, 19, 595–603, 2003.

Ezeji, T.C., Qureshi, N. and Blaschek H.P., Acetone-butanol-ethanol production from concentrated substrate: reduction in substrate inhibition by fed-batch technique and product inhibition by gas stripping, *Appl. Microbiol. Biotechnol.*, 63, 653–658, 2004.

Ezeji, T.C., Qureshi, N. and Blaschek, H.P., Toward acetone butanol (AB) fermentation commercialization: Use of liquefied corn starch and corn steep liquor (CSL) and *in situ* removal by gas stripping, *Biosc. Biotech. Biochem.*, submitted.

Fadeev, A.G., Selinskaya, Y.A., Kelley, S.S., Meagher, M.M., Litvinova, E.G., Khotimsky, V.S. and Volkov, V.V., Extraction of butanol from aqueous solutions by pervaporation through poly(1-trimethylsilyl-1-propyne), *J. Memb. Sci.*, 186, 205–217, 2001.

Fond, O. and Engasser, J.-M., The acetone butanol fermentation on glucose and xylose. 1. Regulation and kinetics in batch cultures, *Biotechnol. Bioeng.*, 28, 160–166, 1986.

Formanek, J., Mackie, R. and Blaschek, H.P., Enhanced butanol production by *Clostridium beijerinckii* BA101 grown in semidefined P2 medium containing 6 percent maltodextrin or glucose, *Appl. Environ. Microbiol.*, 63, 2306–2310, 1997.

Groot, W.J., Soedjak, H.S., Donck, P.B., van der Lans, R.G.J.M. and Luyben, K.C.A.M., Butanol recovery from fermentations by liquid-liquid extraction and membrane solvent extraction, *Bioprocess Eng.*, 5, 203–216, 1990.

Groot, W.J., van der Lans, R.G.J.M. and Luyben, K.Ch.A.M., Batch and continuous butanol fermentation with free cells: integration with product recovery by gas stripping, *Appl. Microbiol. Biotechnol.*, 32, 305–308, 1989.

Groot, W.J., van der Lans, R.G.J.M. and Luyben K.Ch.A.M., Technologies for butanol recovery integrated with fermentations, *Process Biochem.*, 27, 61–75, 1992.

Groot, W.J., van den Oever, C.E. and Kossen, N.W.F., Pervaporation for simultaneous product recovery in the butanol/isopropanol batch fermentation, *Biotechnol. Lett.*, 6, 709–714, 1984.

Harris, L.M., Desai, R.P., Welker, N.E. and Papoutsakis, E.T., Characterization of recombinant strains of the *Clostridium acetobutylicum* butyrate kinase inactivation mutant: Need for new phenomenological models for solventogenesis and butanol inhibition?, *Biotechnol. Bioeng.*, 67, 1–11, 2000.

Huang, W.-C., Ramey, D.E. and Yang, S.-T., Continuous production of butanol by *Clostridium acetobutylicum* immobilized in a fibrous bed reactor, *Appl. Biochem. Biotechnol.*, 113, 887–898, 2004.

Hull, S., Yang, B.Y., Venzke, D., Kulhavy, K. and Montgomery, R., Composition of corn steep water during steeping, *J. Agric. Food Chem.*, 44, 1857–1863, 1996.

Jensen, B.F. and Norman, B.E., *Bacillus acidopullulyticus* pullulanase: application and regulatory aspects for use in the food industry, *Process Biochem.*, 19, 351–369, 1984.

Jones, D.T. and Woods, D.R., Acetone-butanol fermentation revisited, *Microbiol. Rev.*, 50, 484–524, 1986.

Ladisch, M.R., Fermentation—derived butanol and scenarios for its uses in energy—related applications, *Enzyme Microb. Technol.*, 13, 280–283, 1991.

Larrayoz, M.A. and Puigjaner, L., Study of butanol extraction through pervaporation in acetobutylic fermentation, *Biotechnol. Bioeng.*, 30, 692–696, 1987.

Leung, J.C.Y. and Wang, D.I.C., Production of acetone and butanol by *Clostridium acetobutylicum* in continuous culture using free cells and immobilized cells, In: *Proc. 2nd World Cong. Chem. Eng.*, 1, 348–352, 1981.

Maddox, I.S., The acetone-Butanol-Ethanol Fermentation: Recent Progress in Technology, *Biotechnol. Genetic Eng. Reviews*, 7, 190–220, 1989.

Maddox, I.S., Qureshi, N. and Roberts-Thompson, K., Production of acetone-butanol-ethanol from concentrated substrates using *Clostridium acetobutylicum* in an integrated fermentation-product removal process, *Process Biochem.*, 30, 209–215, 1995.

Marchal, R., Rebeller, M. and Vandecasteele, J.P., Direct bioconversion of alkali-pretreated straw using simultaneous enzymatic hydrolysis and acetone butanol production, *Biotechnol. Lett.*, 6, 523–528, 1984.

Marlatt, J.A., Datta, R. Acetone-butanol fermentation process development and economic evaluation, *Biotechnol. Prog.*, 2, 23–28, 1986.

Matsumura, M., Kataoka, H., Sueki, M. and Araki, K., Energy saving effect of pervaporation using oleyl alcohol liquid membrane in butanol purification, *Bioproc. Eng.*, 3, 93–100, 1988.

Mitchell, W.J., Biology and Physiology, In *Clostridia: Biotechnology and Medical Applications*, Bahl, H., Durre, P., Eds., Wiley, Weinheim, Germany, 49–104, 2001.

Mulchandani, A. and Volesky, B., Production of acetone-butanol-ethanol by *Clostridium acetobutylicum* using a spin filter perfusion bioreactor, *J. Biotechnol.*, 34, 51–60, 1994.

Paquet, V., Croux, C., Goma, G. and Soucaille, P., Purification and characterization of the extracellular α-amylase from *Clostridium acetobutylicum* ATCC 824, *Appl. Environ. Microbiol.*, 57, 212–218, 1991.

Parekh, S.R., Parekh, R.S. and Wayman, M., Ethanol and butanol production by fermentation of enzymatically saccharified SO_2-prehydrolyzed lignocellulosics, *Enzym. Microb. Technol.*, 10, 660–668, 1988.

Paturau, J.M., By-products of cane sugar industry: An introduction to their industrialized utilization, 3rd ed., Elsevier, New York, 1989, pp. 265–285.

Pierrot, P., Fick, M. and Engasser, JM., Continuous acetone-butanol fermentation with high productivity by cell ultrafiltration and recycling, *Biotechnol. Lett.*, 8, 253–256, 1986.

Qureshi, N. and Blaschek, H.P., Production of acetone-butanol-ethanol (ABE) by a hyper-butanol producing mutant strain of *Clostridium beijerinckii* BA101 and recovery by pervaporation, *Biotechnol. Prog.*, 15, 594–602, 1999a.

Qureshi, N. and Blaschek, H.P., Butanol recovery from model solution/fermentation broth by pervaporation: Evaluation of membrane performance, *Biomass Bioenergy*, 17, 175–184, 1999b.

Qureshi, N., and Blaschek, H.P., Economics of butanol fermentation using hyper-butanol producing *Clostridium beijerinckii* BA101, Trans. Institution Chemical Engineers (Trans IChemE), (Chemical Engineering Research and Design), 78, 139–144, 2000a.

Qureshi, N., and Blaschek, H.P., Butanol production using hyper-butanol producing mutant strain of *Clostridium beijerinckii* BA101 and recovery by pervaporation, *Appl. Biochem. Biotechnol.*, 84, 225–235, 2000b.

Qureshi, N., and Blaschek, H.P., Evaluation of recent advances in butanol fermentation, upstream and downstream processing, *Bioproc. Biosyst. Eng.*, 24, 219–226, 2001a.

Qureshi, N., and Blaschek, H.P., Recovery of butanol from fermentation broth by gas stripping, *Renewable Energy: An International Journal*, 22, 557–564, 2001b.

Qureshi, N. and Blaschck, H.P., Butanol production from agricultural biomass, In *Advances in Food Biotechnology*, Shetty, K., Pometo, A. and Paliyath, G., Eds, Taylor & Francis, New York, 2005 (in press).

Qureshi, N. and Maddox, I.S., Continuous solvent production from whey permeate using cells of *Clostridium acetobutylicum* immobilized by adsorption onto bonechar, *Enz. Microbial Technol.*, 9, 668–671, 1987.

Qureshi, N., and Maddox, I.S., Reactor design for the ABE fermentation using cells of *Clostridium acetobutylicum* immobilized by adsorption onto bonechar, *Bioproc Eng.*, 3, 69–72, 1988.

Qureshi, N. and Maddox, I.S., Integration of continuous production and recovery of solvents from whey permeate: Use of immobilized cells of *Clostridium acetobutylicum* in fluidized bed bioreactor coupled with gas stripping, *Bioproc. Eng.*, 6, 63–69, 1991a.

Qureshi, N. and Maddox, I.S., An economic assessment of the Acetone-Butanol-Ethanol (ABE) fermentation process, *Australian J. Biotechnol.*, 5, 56–59, 1991b.

Qureshi, N. and Maddox, I.S., Application of novel technology to the ABE fermentation process: An economic analysis, *Appl. Biochem. Biotechnol.*, 34, 441–448, 1992.

Qureshi, N. and Maddox, I.S., Reduction in butanol inhibition by perstraction: Utilization of concentrated lactose/whey permeate by *Clostridium acetobutylicum* to enhance butanol fermentation economics, Official Journal of the European Federation of Chemical Engineering: Food and Bioproducts Processing, Part C, 2005 (in press).

Qureshi, N., Maddox, I.S. and Friedl, A., Application of continuous substrate feeding to the ABE fermentation: Relief of product inhibition using extraction, perstraction, stripping and pervaporation, *Biotechnol. Prog.*, 8, 382–390, 1992.

Qureshi, N., Meagher, M.M., Huang, J. and Hutkins, R.W., Acetone butanol ethanol (ABE) recovery by pervaporation using silicalite-silicone composite membrane from fed-batch reactor of *Clostridium acetobutylicum*, *J. Memb. Sci.*, 187, 93–102, 2001.

Qureshi, N., Meagher, M.M. and Hutkins, R.W., Recovery of butanol from model solutions and fermentation broth using a silicalite/silicone membrane, *J. Memb. Sci.*, 158, 115–125, 1999.

Qureshi, N., Schripsema, J., Lienhardt, J. and Blaschek, H.P., Continuous solvent production by *Clostridium beijerinckii* BA101 immobilized by adsorption onto brick, *World J. Microbiol. Biotechnol.*, 16, 377–382, 2000.

Sedlak, M. and Ho, N.W.Y., Production of ethanol from cellulosic biomass hydrolysates using genetically engineered Saccharomyces yeast capable of cofermenting glucose and xylose, *Appl. Biochem. Biotechnol.*, 113–116, 403–416, 2004.

Sherman, P.D., *Kirk-Othmer Encyclopedia of Chemical Technology*, 3rd ed., John Wiley, New York, 1979, Vol. 4, pp. 338–345.

Singh, V., Johnston, D.B., Moreau, R.A., Hicks, K.B., Dien, B.S. and Bothast, R.J., Pretreatment of wet-milled corn fiber to improve recovery of corn fiber oil and phytosterols, *Cereal Chem.*, 80, 118–122, 2003.

Singh, A. and Mishra, P., Microbial production of acetone and butanol, In *Microbial Pentose Utilization: Current Applications in Biotechnology*, Elsevier Science, New York, 1995, Vol. 33, pp. 119–220.

Soni, B.K., Das, K. and Ghose, T.K., Bioconversion of agro-wastes into acetone butanol, *Biotechnol. Lett. 4*, 19–22, 1982.

Spreinat, A. and Antranikian, G., Analysis of the amylolytic enzymes system of *C. thermosulfurogenes* EM1: purification and synergistic action of pullulanases and maltohexaose forming α-amylase, Starch 44, 305–312. U.S. Department of Energy: Office of Energy Efficiency and Renewable Energy, Office of the Biomass Program, Multiyear Plan 2004 and Beyond, Washington, DC, 2003, 1–40.

Yarovenko, V.L., Principles of the continuous alcohol and butanol-acetone fermentation processes, In *Continuous Cultivation of Microorganisms: 2nd Symposium*, Malek, I., Ed, Czechoslovakian Academy of Sciences, Prague, 1964, pp. 205–217.

Section II

Blended Fuels

7 Ethanol Blends: E10 and E-Diesel

Shelley D. Minteer
Department of Chemistry, Saint Louis University,
Missouri

CONTENTS

Abstract Ethanol was first used as a fuel for a combustion engine in 1897, but it has not emerged into the fuel market as a fuel, but rather as an oxygenate. The United States, Canada, Brazil, and many other countries have adopted the use of ethanol blended with gasoline at low concentration to improve emissions. Ethanol can also be blended with diesel at low concentrations (10% to 15%). Most commonly, ethanol is blended with gasoline at concentration of 10% and this oxygenated fuel is referred to as E10 or gasohol.

HISTORY OF ETHANOL-BASED FUELS

The use of ethanol in an internal combustion engine was first investigated in 1897 (1). Henry Ford originally designed the Model T in 1908 to run on ethanol, but increasing taxes limited its use (2). The concept of employing ethanol as a fuel was reintroduced during the fuel shortages during both World Wars, but the U.S. federal ethanol program was not started until the oil crisis of the 1970s (2). In 1973, OPEC quadrupled the cost of purchasing crude oil (3), which started the resurgence of promoting ethanol as an alternative fuel for combustion engines. However, ethanol as an alternative fuel has not infiltrated the fuel market in the way blended ethanol/gasoline fuels have for automobiles.

Although research in the United States from the Society of Automotive Engineers showed extensive engine testing of E10 (10% ethanol/90% gasoline) in 1933, it was not until 1978 that the U.S. government established a National Alcohol Fuel Commission (4). In 1980, President Carter signed into the law the Energy Security Act containing Title 11, which is commonly called the Biomass Energy and Alcohol Fuels Act of 1980 (4). The Clean Air Act of 1970 allowed the Environmental Protection Agency (EPA) to set standards for vehicle emissions of carbon monoxide, nitrogen oxides, and ozone (4). In 1992, the EPA started requiring cities that were considered to have serious or moderate carbon monoxide pollution problems to establish oxygenated fuel programs. The oxygen content of 2.7% by wt is a required minimum for gasoline sold in these cities. This corresponds to approximately 7.5% by volume ethanol and approximately 15.0% by volume methyl tertiary butyl ether (MTBE) in gasoline.

In 1994, the EPA proposed a policy that at least 30% of the oxygenate be derived from renewable resources (4). However, this proposed policy was not passed by Congress. Ninety-five percent of the oxygenate used in Chicago is ethanol (4). Ethanol has been marketed in every state except California (MTBE has been the mandated oxygenate) (4), but currently MTBE is being phased out of California and ethanol is being phased in due to environmental issues.

In view of the recent Kyoto Conference at which the United States committed to decreasing greenhouse gas emissions by 2012 to below the 1990 level (5), ethanol/gasoline blends from E10 to E85 are an excellent way to achieve these greenhouse gas reductions. Argonne National Laboratory has shown that greenhouse gas emissions is 2.4 to 2.9% less for E10 than 100% gasoline overall (5). Most of this decrease is due to a decrease in greenhouse gas emissions from vehicle combustion because there is actually a small increase in greenhouse gas emission from the fuel due to volatility.

OXYGENATED FUELS

Over the last 30 years, ethanol has been used widely to blend with gasoline in the United States, Brazil, and other countries. In the United States, ethanol is usually blended in a mix of 10% ethanol and 90% gasoline. Early in the use of ethanol blends, this blend was referred to as gasohol, but it is now commonly referred to as E10. The purpose of blending a small percentage of ethanol into gasoline is to oxygenate the fuel for cleaner combustion and fewer carbon monoxide and hydrocarbon exhaust emissions. The most common additive to gasoline to improve oxygen is methyl tertiary butyl ether (MTBE), but it is an extremely toxic chemical that has been found to contaminate groundwater. A comparison of the emission of sulphur, olefins, carbon dioxide, aromatics, and NOx from MTBE oxygenated gasoline and ethanol oxygenated gasoline is shown in Table 7.1. The U.S. Environmental Protection Agency is beginning the process of eliminating MTBE from gasoline (6). Iowa and South Dakota have already phased out MTBE (2). If MTBE were completely replaced with ethanol, it would produce a 12-billion-gallon market for ethanol each year (2), which is considerably more

TABLE 7.1
Comparison of Vehicle Emissions from Using 6% Ethanol/94%
Gasoline and 11% MTBE/89% Gasoline

	Sulfur (ppm)	Olefins (vol%)	CO_2 (mg/gallon burned)	Aromatics (vol%)	RVP (psi)	NOx (g/gallon burned)
Ethanol (6%)	1.22	0.21	8.56	28	6.88	7.4
MTBE (11%)	29.2	3.60	8.74	24	6.71	6.4

Source: Mahy, H., Szabo, C. and Woods, L., *200 Proof Transportation: The Potential for Ethanol as an Alternative Fuel*, University of Washington, Global Commercialization of Environmental Technologies, ENVIR 550/BBUS 550.

ethanol than is currently produced in the United States. To be considered an oxygenated gasoline, the fuel must contain at least 2.7% oxygen by weight. This can be obtained by blending 15% by volume MTBE or 7.5% by volume ethanol, but there is a difference in emissions between the two as shown below (2). Ethanol produces dramatically fewer sulfur and olefin emissions, but comparable emissions of other environmental hazards.

It is important to note most countries do not blend the minimum amount (7.5%) of ethanol for use as an oxygenate. Each country has its own concentration of ethanol to blend with gasoline. E10 has been the choice in the United States. It has also been the choice in areas of Canada. From 1929 to 1957, E10 was the only type of gasoline sold in Queensland. In 2001, E10 was reintroduced to Queensland by the government (7). On the other hand, all gasoline in Brazil is 22% ethanol (E22) (7). Finland has shown that E15 (15% ethanol/85% gasoline) vehicles can operate with stock engines (8). Other countries have considered or employed variations in ethanol concentration from E10 to E25. Table 7.2 below shows how relative emissions change as a function of ethanol concentration. It is important to note that carbon monoxide, hydrocarbon, and NOx emissions

TABLE 7.2
Relative Emissions (% Compared to 100%
Gasoline) as a Function of Ethanol Concentration

Ethanol %	CO	HC	NOx	Aldehydes
100%	29	71	86	1000
95%	36	79	86	Unknown
24%	50	87	120	360
12%	81	96	92	Unknown

Source: Faiz, A., Weaver, C.S. and Walsh, P., *Air Pollution from Motor Vehicles, Standards and Technologies for Controlling Emission*, The World Bank, 1996.

decrease with increasing ethanol concentration, but aldehyde emissions increase with ethanol concentrations. Also, Thailand has shown that the emission rates of benzene, toluene, and xylene are decreased in cars using E10 and E15 fuels (9). This decrease in emissions is important due to the major health effects (including leukemia) of long-term inhalation of benzene and toluene (10). However, E10 and E15 fuels show an increase in formaldehyde and acetaldehyde emissions and exposure to formaldehyde and acetaldehyde has been shown to cause eye irritation, respiratory problems, and nervous disorders (9).

It is also important to consider that E10 is considered an oxygenated fuel, but not an alternative fuel. E85, E95, and biodiesel have large enough biofuel concentrations to be considered alternative fuels, but E10 is simply considered an oxygenated fuel. From 1992 to 1998, the U.S. consumption of vehicle fuel increased by 14.5% (11). However, the U.S. consumption of alternative fuels increased 49.1% and the U.S. consumption of oxygenated fuels has increased 96.9% (11). This shows that more consumers are using alternative and oxygenated fuels today than in 1992. However, there has only been a 21.6% increase in the use of ethanol as an oxygenate (12). This will likely increase as MTBE is phased out due to environmental issues.

Ethanol is an easy fuel to work with because it is liquid at room temperature, can be stored in conventional fuel tanks, is less toxic than many fuels, and is easy to splash blend with gasoline at any stage of the production/distribution process.

ETHANOL PRODUCTION

In 2000, 29.9 billion liters of ethanol were produced worldwide (13). The majority of the production comes from Brazil and the United States. In 2003, 2.8 billion gallons of ethanol were produced in the United States alone (2). Production in 2005 is expected to be approximately 4.0 billion gallons of ethanol (8). The top four producers of ethanol are Iowa (575 million gallons per year), Illinois (523 million gallons per year), Minnesota (486 million gallons per year), and Nebraska (454 million gallons per year). These four states produce approximately 72% of the total ethanol for the United States. The demand for ethanol is approximately divided into 68% fuel, 21% industry, and 11% food and beverages (3). Over 95% of the fuel ethanol produced in the United States was used to make E10, however, a small portion is used for the ever-increasing E85 market.

In comparison, 120 billion gallons of gasoline are sold in the United States each year (2) while only 2.8 billion gallons of ethanol are produced, so the United States does not produce enough ethanol for all gasoline sold to be E10 (maximum ethanol concentrations allowed by the U.S. Environmental Protection Agency) (4). Currently, E10 represents 8% to 10% of the total gasoline sales in the United States (4). This ethanol production shortage is likely to be a major problem as MTBE is phased out and there is more demand for ethanol as an oxygenate. This increase in demand will likely result in a dramatic increase in production of ethanol in the United States.

Ethanol is a controversial fuel. The Renewable Fuels Association states that the ethanol fuel market adds $4.5 billion to farm revenue yearly, employs almost 200,000 people, and increases state tax revenue by $450 million (14). In the United States, there are four federal tax incentives for ethanol sold for fuel: (1) excise tax exemption, (2) blender's tax credit, (3) income tax credit for businesses producing or selling ethanol, and (4) small-producers tax credit for farm co-ops. The first tax benefit is $0.52 per gallon, but the fourth tax benefit is only $0.10 per gallon. Over 30 states have also implemented tax incentives for ethanol as fuel. Most range from $0.20 to $0.40 per gallon. Although many argue assumptions and data, researchers at Cornell University have calculated that a gallon of ethanol requires 29% more energy to produce than it contains as fuel (15). It has also been argued that ethanol production increases environmental degradation, because corn causes more soil erosion than any other farm crop (15). Although soil erosion is an issue, the environmental impacts of ethanol are considerably less than the toxic MTBE. The latest results from the U.S. Department of Agriculture contradict researchers at Cornell University and show that corn ethanol is energy efficient and contains 34% more energy than is required to produce ethanol (16). Part of this dramatic increase in energy efficiency is due to lower energy use in the fertilizer industry and advances in fuel conversion technology over the last decade (16). Similar energy efficiency data has been shown by several other researchers (16–19).

Cost of production of ethanol is a function of plant location, feedstock, production scale, and end use. The choice of feedstock depends on the country. Brazil has used sugar cane as their primary feedstock. France has attempted to use Jerusalem artichokes, but later found that sugar beets and wheat were better for ethanol production. Sweden uses its surplus of wheat to produce the ethanol for their 6% ethanol-blended gasoline. However, in the United States, corn has been determined to be one of the best feedstocks. Approximately, 2.5 gallons of ethanol are produced from every bushel of corn (16), but the corn yield per acre varies as a function of state, along with the fertilizer and irrigation needs in that region. In a corn-based ethanol industry, the cost of the corn is approximately 50–60% of the cost of production of the ethanol (20). It is predicted that the cost of production of ethanol will decrease by $0.11 per liter over the next 10 years due to genetic engineering (21). Currently, the cost of production of ethanol from corn is $0.88 per gallon versus $1.50 gallon from cellulose-based biomass (22). As gas prices rise, the cost of ethanol and ethanol blends becomes more competitive with gasoline.

ENGINE ISSUES

The major engine operation issues with alcohol-blended fuels are fuel quality, volatility, octane number, enleanment, cold start, hot operation, and fuel consumption. The physical properties of the blended fuels compared to pure gasoline are shown in Table 7.3. Octane numbers determined by the usual ASTM procedures indicate that alcohol-blended gasoline increases fuel octane over the base

TABLE 7.3
Physical Properties of Blended Fuels

Physical Property	Gasoline	Ethanol	E10	E20
Specific gravity @ 15.5°C	0.72–0.75	0.79	0.73–0.76	0.74–0.77
Heating value (BTU/gallon)	117,000	76,000	112,900	109,000
Reid vapor pressure @37.8°C (kPa)	59.5	17	64	63.4
Stoichiometric air/fuel ratio	14.6	9	14	13.5
Oxygen content (%wt.)	0	35	3.5	7.0

Source: Guerrieri, D.A., Caffrey, P.J. and Rao, V., *Investigation into the Vehicle Exhaust Emissions of High Percentage Ethanol Blends*, SAE Technical Paper Series, #950777, 1995.

TABLE 7.4
Fuel Economy Decreases with Ethanol Concentration

Ethanol Percentage	Heat of Combustion (BTU/gallon)	Fuel Economy (mpg)
0	115,650	22.00
10	112,080	21.25
14	110,500	20.90
20	108,550	20.48
25	106,510	20.13
30	104,860	20.00
35	102,750	19.57

Source: Alternate Fuels Committee of the Engine Manufacturers Association, *A Technical Assessment of Alcohol Fuels*, SAE 82026. Report to Environment Australia, *A Literature Review Based Assessment on the Impacts of a 10% and 20% Ethanol Gasoline Fuel Blend on Non-Automotive Engines*, Orbital Engine Company, 2002.

gasoline (23–28). Fuel consumption increases when oxygenates are blended with gasoline due to the lower energy content of the oxygenated fuel. Table 7.4 shows that fuel economy decreases with ethanol concentration. The theoretical increase in fuel consuption is 3% for E10 and 6% for E20 (29).

Corrosion of metallic fuel system components is generally not an issue with E10 (28). Researchers have also shown that E20 blends do not appear to affect fuel-system operation (8). Elastomeric and plastic components of new engines are compatible with E10, but many older engines are not (28). Evidence reported has shown that ethanol blends offer less lubrication than pure gasoline (29);

however, that has not been a noticeable issue in terms of wear or engine life over the last 20 years. Over the last few years, Brazil has shown that conventional catalysts used in U.S. vehicles can operate on 10% and neat (100% ethanol) (30). Ref. 8 states that higher ethanol blends show higher catalytic efficiency, because there is a smaller concentration of sulfur species. Barnes (1999), from Ref. 8, says that the increase in catalytic efficiency could be as large as 24%. Guerrieri, Caffrey, and Rao 1995 from Ref. 8 show that volatility decreases with higher ethanol blends. The highest volatility is around 5% ethanol (31). Carbon monoxide emissions are lower for ethanol blends (32–35). E10 can be employed in vehicles without equipment changes and without violating manufacturer's warranties (4).

Enleanment is defined as an excess of oxygen compared to the ideal air/fuel ratio. Common problems of enleanment are loss of power and engine misfires (8). Both problems increase emissions. Finland has shown that E15 vehicles can operate with stock carburetors and that 80% of vehicles running on E15 show less wear compared to pure gasoline. Fluorinated polymers have good resistance to both gasoline and ethanol (36). Nylon-coated nitrile rubber has also shown resistance to both gasoline and ethanol (37). Overall, engine operation and life (wear) are not affected by small (10–20%) concentration of ethanol blended with unleaded gasoline.

E-DIESEL

Since the 1980s, there has been increased interest in low concentration blending of ethanol with diesel fuel. Ethanol/diesel blends are commonly referred to as E-diesel. They generally contain from 10% to 15% ethanol and are used for many of the same reasons that ethanol/gasoline blends are used (decreased petroleum need and decreased emissions). Ethanol and diesel blending is more complicated than ethanol/gasoline blending, because of the low solubility of ethanol in diesel at low temperatures and the high flammability. At temperatures below 10°C, ethanol and diesel will separate [39]. The solution is either to add an emulsifier or a cosolvent. Boruff et al. has shown that approximately 2% surfactant (emulsifier) is needed for every 5% of ethanol added to diesel fuel (40). The addition of the surfactant to the ethanol/diesel blend led to transparent solutions with no visible separation down to −15.5°C (40). Ethyl acetate has been studied as a cosolvent. Researchers have shown that adding 2.5% ethyl acetate for every 5% ethanol will ensure no separation down to 0°C (41). Cosolvents have been more popular than surfactants. The second issue with e-diesel is the increased risk of fire and explosions compared to plain diesel fuel. The National Renewable Energy Laboratory recommends solving this problem by equipping all fuel tanks with vents, better electrical grounding, and employing safer fuel tank level detectors (42). The physical properties of E-diesel compared to ethanol and diesel are shown in Table 7.5.

Blending ethanol with diesel fuel decreases emissions in a similar way to ethanol/gasoline blends. E-diesel has achieved reported 20% to 30% decreases

TABLE 7.5
Physical Properties of E-Diesel

Physical Property	Diesel	Ethanol	E-Diesel (15%)
Vapor pressure @ 37.8C (kPa)	3	15	15
Flashpoint (°C)	64	13	13
Flammability limits			
(%)	0.6 to 5.6	3.3 to 19.0	3.3 to 19.0
(°C)	64 to 150	13 to 42	13 to 42
Density (g/mL)	0.86	0.79	0.85
Heating value (BTU/gallon)	132,000	76,000	123,000

Source: Hansen, A.C., Lye, P.W., Zhang, Q., *Ethanol-diesel blends: A step towards a bio-based fuel for diesel engines,* ASAE Paper No. 01-6048, August 2001; Waterland, L.R., Venkatesh, S., Unnasch, S., *Safety and performance assessment of ethanol/diesel blends (E-Diesel)*, NREL/SR-540-34817, September 2003.

in carbon monoxide emissions and 20% to 40% decreases in particulate matter emissions (43). Miyamoto et al. showed that these improvements in emissions depend directly on the oxygen content (44). However, minimal decreases in NOx emissions have been reported (43) and an increase in hydrocarbon emissions have been reported (45–46). Table 7.6 shows the vehicle emissions from the use of 10% and 15% ethanol in diesel.

As far as engine use is concerned, the decrease in fuel viscosity and lubricity have been investigated for ethanol blends with diesel, but they do meet diesel specifications (42). Materials compatibility has also been investigated. E-diesel was found to have similar corrosive properties to typical diesel (42).

It is important to note that E-diesel fleet demonstrations have shown no fire or explosions incidents and no mechanical failures associated with the fuel system 43). Many studies of engine wear have been conducted and have shown no abnormal wear or deterioration due to the blending of ethanol with diesel at low concentrations (10–15%). E-diesel does shows a reduction in engine power, but this reduction is small and equivalent to the reduction in energy content of the ethanol versus diesel (39). The main engine performance issue with E-diesel is the leakage of fuel from the fuel injection pump due to slight decrease in viscosity of the blended fuel. Studies of engine power loss have shown decreases in power from 4% to 10% for ethanol/diesel blends ranging from 10% to 15% ethanol (46–48). Therefore, ethanol is a good choice as an oxygenate for diesel. It has minimal effect on engine power while dramatically decreasing particulate matter and carbon monoxide emissions.

TABLE 7.6
Vehicle Emissions from the Use of 10% to 15% Ethanol in Diesel

Emissions Range (% ratio of blend/diesel)	10% Ethanol	15% Ethanol
Particulate matter	73–80	59–70
NOx	96–100	95–100
CO	80–160	73–140
Hydrocarbons	171–200	175–210

Source: Data compiled from Hansen, A.C., Lye, P.W. and Zhang, Q., ASAE Paper No. 01-6048, Aug. 2001; Waterland, L.R., Venkatesh, S. and Unnasch, S., NREL/SR-540-34817, Sept. 2003; Spreen, K., Final Report for Pure Energy Corporation prepared at SRI, San Antonio, TX, 1999; Kass, M.D., Thomas, J.F., Storey, J.M., et al., SAE Technical Paper 2001-01-2018, 2001.

CONCLUSIONS

Ethanol can be blended with gasoline to produce an oxygenated fuel with lower hydrocarbon emissions. Ethanol can also be blended with diesel to decrease carbon monoxide emissions and particulate matter emissions. Although greenhouse gas emissions are decreased with ethanol-blended fuels, emissions of certain aldehydes are increased, which could cause health issues. Automobiles can be operated on ethanol/gasoline blends from 5% to 25% and ethanol/diesel blends from 10% to 15% without need for any alterations in engine equipment or settings and with no effect on engine lifetime.

REFERENCES

1. Rothman, H., Greenshields R. and Calle F.R., *The Alcohol Economy: Fuel Ethanol and the Brazilian Experience*, Francis Printer, London, 1983.
2. Mahy, H., Szabo, C. and Woods, L., *200 Proof Transportation: The Potential for Ethanol as an Alternative Fuel*, University of Washington, Global commercialization of Environmental Technologies, ENVIR 550/BBUS 550, Hund, G. and Laverty, K., Eds., 2003.
3. Taherzadeh, M.J., *Ethanol from Lignocellulose: Physiological Effects of Inhibitors and Fermentation Strategies*, Chalmers University of Techology, Dept. of Chemical Engineering, Goteborg, Sweden, 1999.

4. Prakash, C., Motor Vehicle Emissions and Fuels Consultant, *Use of Higher than 10 Volume Percent Ethanol/Gasoline Blends in Gasoline Powered Vehicles.* Transportation Systems Branch Air Pollution Prevention Directorate Environment Canada, 1998.

5. Wang, M., Saricks, C., Wu, M., *Fuel-Cycle Fossil Energy Use and Greenhouse Gas Emissions of Fuel Ethanol*, Illinois Department of Commerce and Community Affairs, Center for Transportation Research, Argonne National Laboratory, 1997.

6. Browner, C., Remarks as prepared for delivery to press conference on March 20, 2000, http://www.epa.gov/otaq/consumer/fuels/mtbe/press34b.pdf. (accessed 2002).

7. Kimble, D., Is ethanol fuel the answer to our environmental problems?

8. (October 2002), *Issues Associated with the Use of Higher Ethanol Blends (E17-E24)*, Hammel-Smith, C., Fang, J., Powders, M. and Aabakken, J., Eds., National Renewable Energy Laboratory, Golden, CO, 2002.

9. Leong, S.T., Muttamara, S. and Laortanakul, P., Applicability of gasoline containing ethanol as Thailand's alternative fuel to curb toxic VOC pollutants from automobile emission, *Atmospheric Environment*, 2002.

10. U.S. EPA, *Motor vehicle-related air toxic study*, U.S. Environmental Protection Agency, Office of Mobile Sources, Ann Arbor, MI, EPA Report No. EPA 420-R-93-005, April 1993.

11. EIA, Energy Information Administration: *Alternative to traditional transportation fuels 1996*, U.S. Department of Energy, DOE/EIA-0484(98), 1998.

12. Faiz, A., Weaver, C.S. and Walsh, M.P., *Air Pollution from Motor Vehicles, Standards and Technologies for Controlling Emission*, The World Bank, 1996, p. 207.

13. Berg, C., *World Ethanol Production* 2001, http://www.distill.com/world_ethanol_production.htm (accessed 2002).

14. Renewable Fuels Association, Official Statement, http://www.ethanol-rfa.org/leg_official.shtml.

15. Pimentel, D., *Ethanol Fuels: Energy Balance, Economics, and Environmental Impacts Are Negative*, Natural resources research, 12, 2, Kluwer Academic/Plenum Publishers, New York, 2003.

16. Shapouri, D., James A. and Wang, M., *The Energy Balance of Corn Ethanol: An Update*, Agricultural Economic Report Number 814, U.S. Department of Agriculture, 2002.

17. Agriculture and Agri-Food Canada, Assessment of Net Emissions of Greenhouse Gases From Ethanol-Gasoline Blends in Southern Ontario, Prepared by Levelton Engineering Ltd. #150-12791, Clarke Place, Richmond, B.C. and (S&T)² Consulting Inc., J.E. & Associates, Aug. 1999.

18. Lorenz, D. and David M., *How Much Energy Does it Take to Make a Gallon of Ethanol?* Revised and Updated, Institute for Local Self-Reliance, Washington, DC, Aug. 1995.

19. Marland, G., and Turhollow, A.F., CO_2 *Emissions from the Production and Combustion of Fuel Ethanol from Corn*, Oak Ridge National Laboratory, Oak Ridge, TN, Environmental Sciences Division, No. 3301. U.S. Department of Energy, May 1990.

20. Kaylen, M., Van Dyne, D.L., Choi, Y.S., and Blasé, M., Economic feasibility of producing ethanol from lignocellulosic feedstocks, *Bioresource Technology*, 72, 2000, 19–32.

21. Wooley, R., Ruth, M., Glassner, D. and Sheehan, J., Process design and costing of bioethanol technology: a tool for determining the status and direction of research and development, *Biotechnology Progress*, 15, 794–803, 1999.

22. McAloon, A., Taylor, F., Yee, W., Ibsen, K. and Wooley, R., *Determining the cost of producing ethanol from corn starch and lignocellulosic feedstocks*, National Renewable Energy Laboratory (NREL), Golden, CO, http://www.afdc.doe.gov/pdfs/4898.pdf (accessed 2002).

23. Alternate Fuels Committee of the Engine Manufacturers Association, *A Technical Assessment of Alcohol Fuels*, SAE 82026.

24. Birrell, J.S., *Ethanol as a petrol extender in spark ignition engines*, SAE 825026.

25. Brinkman, N.D. et al, *Exhaust emissions, fuel economy and driveability of vehicles fuelled with alcohol-gasoline blends*, SAE 750120.

26. Owen, K. and Coley, T., *Automotive Fuel Handbook*, 1990.

27. Wagner, T.O. et al, *Practicality of alcohols as motor fuels*, SAE 790429.

28. Report to Environment Australia, *A Literature Review Based Assessment on the Impacts of a 10% and 20% Ethanol Gasoline Fuel Blend on Non-Automotive Engines*, Orbital Engine Company, 2002.

29. Black, F., *An overview of the technical implications of methanol and ethanol as highway motor vehicle fuels*, SAE 912413.

30. Szwarc, A., ADS Technology and Sustainable Development, State of California (Mar. 1999), Governor Gray Davis, Executive Department, Executive Order D-5-99, e-mail, July 26, 1999.

31. American Petroleum Institute, *Alcohol and Ethers: A Technical Assessment of Their Application as Fuels and Fuel Components*, API Publication 4261, third ed., June 2001.

32. Myron, W.B. et. al., *Ethanol Blended Fuel as a CO Reduction Strategy at High Altitude*, Colorado Department of Health, Air Pollution Control Division, Aug. 1985.

33. California Air Resources Board, http://www.arb.ca.gov.

34. Guerrieri, D.A., Caffrey, P.J. and Rao, V., *Investigation into the Vehicle Exhaust Emissions of High Percentage Ethanol Blends*, SAE Technical Paper Series, #950777, 1995.

35. Faiz, A., Weaver, C.S. and Walsh, M.P., *Air Pollution from Motor Vehicles, Standards and Technologies for Controlling Emission*, The World Bank, 1996.

36. Nersasian, A., (E.I. Du Pont de Nemours & Co., Inc.), *The Vol. Increase of Fuel Handling Rubbers in Gasoline/Alcohol Blends*, SAE Paper No. 800789, Jun. 1980.

37. Dunn, J.R. and Pfisterer, H.A., (Polystar Limited), *Resistance of NBR-Based Fuel Hose Tube to Fuel-Alcohol Blends*, SAE Paper No. 800856, Jun. 1980.

38. Environment Australia, Setting the ethanol limit in petrol.

39. Hansen, A.C., Lye, P.W. and Zhang, Q., *Ethanol-diesel blends: A step towards a bio-based fuel for diesel engines*, ASAE Paper No. 01-6048, Aug. 2001.

40. Boruff, P.A., Schwab, A.W., Goering, C.E. and Pryde, E.H., Evaluation of diesel fuel-ethanol microemulsions, *Transactions of ASAE*, 25, 47–53, 1982.

41. Letcher, T.M., Diesel blends for diesel engines, *South African J. Sci.*, 79(1), 4–7, 1983.

42. Waterland, L.R., Venkatesh, S. and Unnasch, S., *Safety and performance assessment of ethanol/diesel blends (E-Diesel)*, NREL/SR-540-34817, Sept. 2003.

43. Marek, N. and Evanoff, J., Pre-commercialization of E-diesel fuels in off-road applications, In *Proceedings of A&WMA 2002 Annual Conference*, Paper No. 42740, June 2002.

44. Miyamoto, N., Ogawa, H., Nurun, N., Obata, K. and Arima, T., *Smokeless, low nox, high thermal efficiency, and low noise diesel combustion with oxygenated agents as main fuel*, SAE Technical Paper No. 980506, 1998.

45. Spreen, K., *Evaluation of oxygenated diesel fuels*, Final Report for Pure Energy Corporation Prepared at SRI, San Antonio, TX, 1999.

46. Kass, M.D., Thomas, J.F., Storey, J.M., Domingo, N., Wade, J. and Kenreck, G., *Emissions from a 5.9 liter diesel engine fueled with ethanol diesel blends*, SAE Technical Paper 2001-01-2018, 2001.

47. Hansen, A.C., Hornbaker, R.H., Zhang, Q. and Lyne, P.W., *On-farm evaluation of diesel fuel oxygenated with ethanol*, ASAE Paper No. 01-6173, 2001.

48. Hansen, A.C., Mendoza, M., Zhang, Q. and Ried, J.R., *Evaluation of oxydiesel as a fuel for direct-injection compression-ignition engines*, Final Report for Illinois Department of Commerce and Community Affairs, Contract IDCCA 96-32434, 2000.

8 Using E85 in Vehicles

Gregory W. Davis, Ph.D., P.E.
Director, Advanced Engine Research Laboratory and
Professor of Mechanical Engineering, Kettering University,
Flint, Michigan

CONTENTS

Abstract The use of E85 as a fuel for vehicles is discussed in this chapter. E85, a blend of 85% ethanol and 15% gasoline is a liquid fuel that can be utilized in a wide variety of vehicles. The use of E85 has been encouraged because it dramatically reduces exhaust and greenhouse gas emissions. Additionally, the ethanol used can be derived from renewable sources. In order to use E85, the vehicle must have a spark-ignited engine. Furthermore, this engine must be adapted to accept E85.

E85 is a high-blend alcohol-based fuel containing 85% ethanol and 15% gasoline by volume. Because pure ethanol has a lower vapor pressure than gasoline, it is blended with 15% gasoline to produce E85 in order to minimize difficulties in starting engines and with drivability during cold weather.

E85 is used to operate spark-ignited engines that have been modified to accept this fuel. Spark-ignited engines cannot directly use E85 without modification due to higher mixture requirements and some material compatibility issues. E85 is used in vehicles to reduce vehicle tailpipe emissions. Further, E85, in which the ethanol has been derived from biomass, also reduces the net production of carbon dioxide, a greenhouse gas. Additionally, E85 is becoming cost competitive with gasoline, with wholesale prices lower than gasoline in early 2005 in the United States. Because of this, and the fact that ethanol can be produced from renewable resources instead of petroleum has led to its development and use in a variety of vehicles. The use of E85 is growing rapidly in the United States; however, the total number of vehicles currently using E85 is still small when compared with the total number of vehicles on the road.[1]

HISTORY

High-blend alcohol fuels have been used in vehicles for many years in different regions of the world. Brazil is probably the most well-known region, having established government policies in the 1970s to develop this fuel.[2] During the 1990s, 4.5 million automobiles operating on 93% ethanol (balance gasoline), were in use in Brazil.[3] In the United States, E85 has received increasing attention and use due to stricter emissions standards and worries about energy security. For example, in 1992 the federal government passed the Energy Policy Act (EPAct).[4] This legislation was established with the goals of enhancing the nation's energy security and improving environmental quality. The EPAct encourages the development and use of alternative fuels that are not substantially derived from petroleum.

Alternative fuels are defined to include alcohols at blends of 85% or more of alcohol (such as ethanol) with gasoline. The U.S. Department of Energy (DoE) is charged with the responsibility of implementing this act. The EPAct contains both voluntary and mandatory provisions designed to develop an alternative fuel economy. The EPAct's voluntary activities are administered through the DoE Clean Cities Program, which helps create markets for alternative fuels and alternative fuel vehicles (AFVs) through public/private partnerships in more than 80 U.S. cities.[5]

The mandatory EPAct provisions consist of four programs: The State & Alternative Fuel Provider Program; The Federal Fleet Program; Alternative Fuel Petitions Program; and the Private & Local Government Fleet Program. These programs give the DoE the power to require Federal and State governmental agencies to purchase AFVs as a percentage of their vehicle acquisitions. The Private & Local Government Fleet Program even gives the DoE the authority to impose AFV acquisition requirements on private and local government fleets, although this program has not been implemented.

In addition to the EPAct, the U.S. Federal government maintains a system of tax incentives for E85 in order to encourage its use and development. For example, a Volumetric Ethanol Excise Tax Credit of 51 cents per gallon of ethanol used in fuel is currently available to transportation fuel producers.[6] This law also eliminates alternative minimum tax (AMT) on the Alcohol Fuels Income Tax Credit. Small ethanol producers are also provided with a 10 cents per gallon tax credit on up to 15 million gallons of production annually. Finally, states and the U.S. Federal government offer many grants to help in the production and use of E85.

In the United States, legislation and incentives have led to the development and use of many E85 capable vehicles. The U.S. Energy Information Administration (EIA) estimates that more than four million Flexible Fueled Vehicles were on U.S. roadways in 2002.[7] The annual growth in E85 capable vehicles from 1996 to 2005 was 78.8%, and the projected E85 fuel use was projected to grow by 11.5% from 2003 to 2004 according to the EIA.[1]

TYPES OF VEHICLES USING E85

E85 is used to operate spark-ignited engines that have been modified to accept this fuel. Unlike E10, E85 cannot be used in spark-ignited engines that have not been modified. This has slowed the adoption of E85 because there is a supply and demand problem: consumers will not buy vehicles unless there is a readily available source of fuel, and fuel companies will not invest in the alternative fuels unless there is a large supply of vehicles that use the fuel. This has led to the development of a new type of vehicle, called a Flexible Fuel Vehicle, which can operate using various blends of alcohol.

FLEXIBLE FUEL VEHICLES

Flexible Fuel Vehicles, or FFVs, can operate on ethanol-blends from 0% (gasoline) up to 85% (E85) by volume. This eliminates the supply and demand problem as consumers can fuel with any combination of fuel, not worrying whether the correct fuel will be available. These vehicles are produced by most of the major automakers and represent the largest class of vehicles using E85. Most of these vehicles are sold without any cost penalty to the consumer. This is both an indication of the automakers' desire to develop the market and the incremental costs required to produce these vehicles compared with their gasoline-fueled counterparts.

These vehicles are designed and manufactured using E85 compatible materials. Further, due to the different fuel-air mixture requirements of gasoline and E85, the fuel delivery systems are sized to handle the increased volumes of fuel when using E85. Other changes are made to the control algorithms in order to optimize the vehicle for use with this fuel. Since E85 has a higher octane than gasoline, the spark timing can be advanced, improving engine performance. Further, the fuel injection pulsewidth or duration must be lengthened to increase the flowrate of E85 for given engine load and speed conditions.

Since consumers can fuel these vehicles with either E85 or gasoline (or any blend in between), these vehicles must determine the levels of ethanol present onboard the vehicle in order to ensure that the engine is operating at the best conditions for the given fuel blend. In order to accomplish this, current FFVs have a fuel sensor, which is located in the fuel delivery lines leading from the fuel tank to the engine. The fuel sensor measures the conductivity of the current fuel blend. Since ethanol and gasoline have vastly different levels of conductivity (ethanol is about 135,000 times more conductive than gasoline),[8] this is a relatively easy task to accomplish with some precision. Older FFVs relied on the use of a feedback signal from an exhaust gas oxygen (EGO) sensor located in the exhaust stream. This sensor, already present on all spark-ignited vehicles, detects the presence of excess oxygen in the exhaust. Because of exhaust after-treatment requirements on-road vehicles generally operate using stoichiometric mixtures of fuel and air. Thus, the oxygen sensor is used to maintain stoichiometric combustion in an engine. This sensor can be used to determine fuel mixture as ethanol is an oxygenated fuel with a richer stoichiometric mixture; thus, the mixture used for gasoline will be too lean with E85 and lead to excess oxygen in the exhaust. Unfortunately, these sensors only function when warm, so they cannot be used to help during cold starts of the engine. Thus, going from gasoline to E85 results in a lean mixture until the EGO is functioning. This can lead to poorer quality cold starts and poor drivability under acceleration.[9]

Many environmental groups have been critical of using FFVs, since modifying gasoline-powered vehicles to operate using E85 puts E85 at an inherent disadvantage and it does not force the rapid buildup of an E85 fueling infrastructure since consumers can continue to use gasoline.[2] It is important to remember that the automakers need to produce vehicles that consumers will actually buy; most consumers will not buy a vehicle for which there are few fueling opportunities. This criticism is perhaps premature at this early stage of E85 development.

OFF-ROAD VEHICLES

Although many E85 vehicle demonstrations have been made using off-road vehicles such as airplanes,[10] snowmobiles,[11] boats, and all-terrain vehicles, there are not currently any significant numbers of these vehicles operating on E85.

MATERIAL COMPATIBILITY

Some materials that are commonly used with gasoline-powered vehicles are not compatible with E85. These materials are degraded when in contact with E85 and cause leaks or fuel system contamination.[13] Fortunately, there are many alternatives for these materials. Also, limited duration contact with E85 in many of these materials has shown no detrimental effects. Most degradation requires long-term contact with E85.

E85 can be used in both four-stroke and two-stroke spark-ignited engines. Four-stroke engines are widely used in on-road vehicles because they generally offer better emissions and fuel consumption than two-stroke spark-ignited engines. In countries with strict air-pollution standards, even most motorcycles generally employ this type of engine. The strict emissions standards are also contributing to more widespread use of four-stroke engines for off-road vehicle use, such as snowmobiles and all-terrain vehicles (ATVs).

Most four-stroke spark-ignited engines currently available today introduce the fuel into the air intake system, not directly into the cylinder. This means that the fuel will come into contact with the materials used in the intake manifold of the engine. Fuel also comes into contact with the engine cylinders and the fuel induction and storage systems of these engines.

Two-stroke engines are lighter and often have better power-to-weight ratios than four-stroke spark-ignited engines and are, therefore, often used in smaller vehicles or in cases where weight is a major design consideration.

Developing countries still widely use two-stroke spark-ignited engines in vehicles due to their lower costs and smaller sizes. Two-stroke spark-ignited engines complete a mechanical cycle in two strokes of the piston, or one engine revolution. These types of engines do not have separate intake and exhaust processes. Because of this, these engines produce power every revolution, leading to smaller, lighter engines. Unfortunately, this also leads to higher tailpipe emissions and problems with bypass, where raw fuel and air pass through the engine unburned. To help combat this, many two-stroke spark-ignited engines use crankcase compression to improve scavenging efficiency and to reduce bypass.

This means that in two-stroke spark-ignited engines fuel not only comes into contact with the fuel storage and delivery systems, the intake, and the engine cylinders, but also the engine crankcase and even the exhaust manifold. Further, residual fuel is left in the crankcase after the engine is stopped, leading to potential long-term exposure.

Unfortunately, to save weight, most of these engines use aluminum extensively in their blocks, leading to potential long-term corrosion problems. As described later, hard-anodized aluminum has been shown to be resistant to E85 degradation. At this time, the long-term use of E85 in off-road vehicles with two-stroke engines has not been studied.

METALLIC SUBSTANCES

Metallic substances that are degraded by E85 include: zinc, brass, aluminum, and lead-plated steel. Alloys containing these metals must be individually investigated to determine their E85 compatibility. For example, lead-tin alloy is not E85 compatible. Unfortunately, many vehicles use aluminum in the fuel delivery systems to save weight, including in the fuel pump, lines, fuel rail, and fuel pressure regulator. Fuel also often is allowed to contact the aluminum block of many two-stroke engines. Furthermore, older vehicles often use lead-plated steel for the vehicle fuel storage tanks. These materials will react with E85, partially dissolving in the fuel. This can contaminate the fuel system, leading to clogged fuel filters and injectors, which in turn, cause poor vehicle drivability. Aluminum can be safely used if it is hard anodized or nickel plated. Most FFVs use hard anodized aluminum for the fuel delivery systems. Also, most modern vehicles use fuel storage tanks that are made of polymer compounds (which are resistant to E85) instead of lead-plated steel; thus, this problem is only a factor in older vehicles.

Other metallic compounds that are resistant to E85 include: unplated steel, stainless steel, black iron, and bronze. These materials can be substituted for the other compounds as required.

Nonmetallic Substances

Nonmetallic materials that degrade when in contact with E85 include natural rubber, polyurethane, cork gasket material, leather, polyvinyl chloride (PVC), polyamides, methyl-methacrylate plastics, and certain thermo and thermoset plastics.

This author has had much experience using E85 in a variety of vehicles with plastic fuel tanks with no noticeable negative consequences. The types of vehicle tanks tested include late model automobiles and light-duty trucks, snowmobiles, small engines, and many plastic fuel delivery tanks. Many of these tanks are made of thermo/thermoset plastics, so this appears not be a major issue for vehicles.

Older vehicles may still use rubber, polyurethane or cork gaskets and O-rings for sealing fuel delivery systems; fortunately, most late model vehicles (vehicles produced after the mid-1990s) no longer use these materials in favor of more advanced sealants.

Many of the other sensitive materials are not used in areas where they might come into contact with the fuel; however, care should be taken to ensure that fuel spillage is cleaned from leather or plastic interior surfaces of the vehicles.

Nonmetallic materials that are resistant to E85 degradation include nonmetallic thermoset reinforced fiberglass, thermoplastic piping, thermoset reinforced fiberglass tanks, Buna-N, Neoprene rubber, polypropylene, nitrile, Viton, and Teflon. All of these materials may be used with E85. Furthermore, most modern vehicles already use these materials for gaskets and O-rings as they offer superior leak resistance. For example, most automakers now use Viton O-rings to seal their fuel injectors.

Vehicle Fuel Pumps

During the mid-1990s, many gasoline fuel pumps suffered high failure rates when delivering E85. Early on, the lower lubricity of ethanol was blamed for these failures. Later, it became clear that the much higher electrical conductivity (ethanol is about 135,000 times more conductive than gasoline) was at least partly to blame. These problems have been addressed by the automakers and premature failures are no longer a problem. These "hardened" pumps are now standard on many vehicles that are not specifically rated for E85 due to their superior performance and reduced failure rates.

PROPERTY COMPARISON WITH GASOLINE

Like gasoline, ethanol is liquid at room temperature and pressure. It can be handled and dispensed using equipment designed for gasoline-with some modifications to accommodate material incompatibilities as discussed above. Most consumers would not notice any difference when fueling their vehicles using E85.

One of the major differences between using E85 and gasoline affecting engine operation is due to the differences in vapor pressure and latent heat of vaporization. In order for combustion to begin in an engine, a portion of the fuel must be vaporized. Gasoline is a mixture of many hydrocarbon compounds with varied vapor pressures and latent heats of vaporization. This means that even under cold conditions a portion of gasoline will still evaporate. Because ethanol is a pure substance, it becomes difficult to vaporize when cold. In fact ethanol will not form an air/fuel vapor mixture high enough to support combustion below 11°C.[12] This led to the use of E85. Gasoline is added to the ethanol in order to support cold startability. Most E85 is blended with regular grade unleaded gasoline.

A comparison of E85 and gasoline is presented in Table 8.1. One complication in using values from this table is the fact that E85 is made from ethanol that has been denatured with up to 5% gasoline; thus E85 is usually composed of less than 85% pure ethanol. This means that the data in Table 8.1 is an approximation of E85 as it is based upon a true blend of 85% ethanol.

Further, E85 is determined on a volume basis, but many users mistakenly use a mass basis in order to determine its composition.[3] Fortunately, the densities of gasoline and E85 are similar as shown in Table 8.1. Assuming constant component volumes during mixing, 85% ethanol on a volume basis produces about 85.7% ethanol on a mass basis.

Finally, the actual blends of E85 are seasonally adjusted depending on the geographical region and the season. During warm weather the blends have higher levels of ethanol to lower vapor pressure; thus minimizing evaporative emissions and vapor lock. These blends of E85 typically contain 85% denatured ethanol. While in cold weather, more gasoline is added to the blend to avoid starting problems. Most winter blends of E85 are actually 70% ethanol by volume. Of course, the gasoline, too, is seasonally adjusted to minimize vapor pressure during the warmer months and to aid in cold startability during the colder months. As one can see, the values in Table 8.1 are simply nominal values. Samples of E85 used in emissions testing should first be analyzed by a qualified laboratory to obtain precise property values.

American Society for Testing and Materials (ASTM) standards for E85 are presented in the following table. These standards, although generally voluntary, are usually followed by major fuel producers. Table 8.2 lists physical properties

TABLE 8.1
Physical Fuel Properties

Physical Property	Gasoline-Regular Unleaded	Ethanol	E85
Formulation	C_4 to C_{12} H/C – chains	C_2H_5OH	85% Ethanol (vol) 15 % Gasoline (vol)
Average Analysis (% mass)	C: 85-88 H: 12–15	C: 52 H: 13 O:35	C: 57 H: 13 O: 30
Octane- (R + M)/2	87	98-100	96
Specific Gravity (60/65 F)	0.72–0.78	0.794	0.785
Lower Heating Value – Btu/lb_m(KJ/Kg)	18,500 (43000)	11,500 (26750)	12,500 (29080)
Lower Heating Value – Btu/gal (KJ/liter)	115,700 (32,250)	76,200 (21,240)	81,900 (22,830)
Reid Vapor Pressure – psi (kPa)	8–15 (50–100)	2.3 (15)	6-12 (41-83)
Heat of Vaporization – Btu/lbm (KJ/Kg)	140–170 (330–400)	362–400 (842–930)	349 (812)
Flammability Limits – % Fuel (volume)	1–8	3–19	–
Stoichiometric A/F (mass)	14.7	9	10
Conductivity – (mhos/cm)	1×10^{-14}	1.35×10^{-9}	–

Source: Data compiled from Davis, G., et al., Society of Automotive Engineers, 1999-01-0609, 1999; U.S. Department of Energy, DOE/GO-1002001-956, Revised Oct. 2002; Society of Automotive Engineers, 930376, 1993.

for the different seasonal blends. Note that this table lists the true levels of pure ethanol; thus the levels appear lower than expected due to the gasoline used as the denaturant in the ethanol.

Class 1 (minimum 79% ethanol) is generally considered summer blend; it is used by most states during the warm months. This is the fuel that is closest to "true" E85. Classes 2 and 3 are considered winter, or spring/fall blends. Class 2 (minimum 74% ethanol) is generally used during spring and fall in cooler climates, and in the winter in mild climates. Class 3 (minimum 70% ethanol) is used during the winter, early spring and late fall, in cooler climates.

One additional property of ethanol that is not shown in either table is its high miscibility with water. Water entrainment in the ethanol can cause the ethanol and gasoline to separate, leading to vehicle stalls and poor drivability.[15] Fuel handling and storage systems must be designed to keep moisture levels out of the fuel. On the positive side, regular use of E85 helps to eliminate moisture in vehicle fuel storage systems as the moisture is entrained in the ethanol and then removed from the system.

TABLE 8.2
ASTM D5798-99 Specification for Seasonal Blends of E85

Physical Property	Value for Class			ASTM Test Method
ASTM volatility	1	2	3	–
Minimum level of ethanol (plus higher alcohols) – % volume	79	74	70	D5501
Hydrocarbons (including denaturant) – % volume	17–21	17–26	17–30	D4815
Vapor pressure (37.8°C) – psi	5.5–8.5	7.0–9.5	9.5–12.0	D4953, D5190,
(kPa)	(38–59)	(48–65)	(66–83)	D5191
Sulfur (maximum) – mg/kg	210	260	300	D3120, D1266, D2622
Water (maximum) – % mass	1.0	1.0	1.0	E203
Acidity (as acetic acid) – ppm	50	50	50	D1613

Source: U.S. Department of Energy DOE/GO-1002001-956, Revised Oct. 2002.

EFFECT OF E85 ON VEHICLE FUEL ECONOMY, PERFORMANCE, AND SAFETY

Table 8.1 compares the physical properties of E85 and gasoline. The property differences that exert the most influence on vehicle performance are: octane, energy density, Reid vapor pressure, stoichiometric A/F mixture, heat of vaporization, and flammability limits. The effects of these different properties are described below.

FUEL ECONOMY

The energy density on a mass basis for E85 is only about 68% of the level for gasoline. Fortunately, the specific gravity of E85 is slightly greater than that for gasoline leading to an E85 energy density of 71% of that for gasoline on a volume basis. Therefore, in order to achieve the same level of power, and assuming no change in engine efficiency, a vehicle operating on E85 would have to consume about 1.4 times as much fuel on a volume basis. This would lead directly to a 29% loss in fuel economy. However, in practice, this reduction is limited as the E85 fuel burns more cleanly, and the engine calibration is adjusted to advance the spark timing, further improving engine efficiency.

Actual test values for FFVs are published by the U.S. federal government; a portion of this is shown in Table 8.3. This data reveals an average loss of about 25% in fuel economy on both the federal highway and city tests when going from gasoline to E85 in FFVs.[16] It is interesting to note that the test data reveal losses as high as 29% and as low as 20%, demonstrating the effect that proper engine

TABLE 8.3
2005 Flexible Fuel Vehicle Federal Fuel Economy Values

Vehicle	Fuel	City Fuel Economy, mpg	Highway Fuel Economy, mpg
Ford Taurus Wagon:	Regular gasoline	19	26
6 cyl, 3 L, Auto(4)	E85	14	19
Mercedes-Benz C320 Sports	Premium gasoline	19	24
Coupe FFV:	E85	14	18
6 cyl, 3.2 L, Auto(5)			
Dodge Caravan 2WD:	Regular gasoline	18	25
6 cyl, 3.3 L, Auto(4)	E85	13	17
Chrysler Voyager/Town & Country	Regular gasoline	18	25
2WD:	E85	13	17
6 cyl, 3.3 L, Auto(4)			
Chevrolet C1500 Silverado 2WD:	Regular gasoline	16	20
8 cyl, 5.3 L, Auto(4)	E85	12	16
GMC C1500 Sierra 2WD:	Regular gasoline	16	20
8 cyl, 5.3 L, Auto(4)	E85	12	16

Source: U.S. Department of Energy, http://www.fueleconomy.90v.

calibrations can have when using E85 with the same FFV. Other sources have suggested lower losses in fuel economy (miles per gallon) of only a 5% to 12% during real-world driving conditions.[13]

A dedicated E85 vehicle could perform better by taking advantage of the higher octane of E85 compared to gasoline. As shown in Table 8.1, E85 enjoys about a 10% advantage in octane rating. Studies have shown that engines could then be designed with higher compression ratios, increasing their operating efficiency by up to 10%.[14] This efficiency, coupled with the increased power extraction during the expansion stroke of the engine due to the increased volume of the combustion products results, in a total efficiency increase of up to 15% compared to gasoline engines. If vehicles were designed to take full advantage of E85, they would probably experience a fuel economy penalty of about 14% on a volume basis when compared with a gasoline powered vehicle. Although it is important to note that engine calibrations, as shown earlier, can have a dramatic impact upon this value.

In conclusion, it is important to note that the actual energy efficiency for vehicles using E85 is higher than those using gasoline; however, the fuel economy, expressed on a miles-per-gallon basis, is lower due to the lower energy density of E85 on a volumetric basis. Thus, it is environmentally beneficial to use E85 even though its use will probably result in higher fuel usage on a volumetric basis.

Vehicle Power

Ethanol has long been used in racing because of its desirable properties for increasing engine power output. E85, too, increases the power and torque capability of engines compared with gasoline. Most spark-ignited engines used in on-road vehicles operate with air to fuel mixtures at or near the stoichiometric condition. Since the stoichiometric air to fuel ratio of E85 is less than gasoline, engines operating on E85 can use about 1.48 times more E85 for the same amount of air. Remembering that about 1.4 times more E85 is required to equal the energy of gasoline on a volume basis, this leads to about a 6–7% increase in power.

Since E85 burns cleaner and the engine spark timing can be advanced due to the increased octane number, an engine operating on E85 can actually achieve higher increases in power.

Additionally, E85 has a higher heat of vaporization than gasoline. This is important in spark-ignited engines as the fuel is inducted into the intake manifold. As the fuel vaporizes due to the heat of the engine, it displaces air, reducing the ability of the engine to draw in fresh air. This reduces the volumetric efficiency of the engine, reducing the power. By increasing the heat of vaporization, E85 increases the engine volumetric efficiency, allowing more air to be drawn into the engine. This additional air allows the engine to use additional fuel, leading to increased power.

Further, an engine can be designed with a higher compression ratio to take full advantage of the increased octane of E85. All of these factors can be combined to increase power by more than 25% compared with the same-sized gasoline engine.

Cold Startability

Pure ethanol becomes difficult to vaporize when cold, leading to poor cold startability. In fact ethanol will not form an air/fuel vapor mixture high enough to support combustion below 11°C.[12] Therefore, gasoline is added to the ethanol in order to support cold startability and increased cold start enrichment is used to achieve combustible vapor air mixtures in the engine.

Additional difficulties when cold starting with E85 can be attributed to its high conductivity. During cold starts, the spark plug electrodes can become wetted with fuel. Since E85 is much more conductive, this can leading to plug shorting and misfire.[8]

These problems have been addressed by the major automakers through better cold-start fuel calibrations. Most manufacturers now report good cold starts at temperatures below 0°F (−18°C) when using E85 in a winter blend (E70).

Safety

E85 has wider flammability limits on the rich side, and it has a higher flame speed compared to gasoline. This increases the probability of encountering flashback or fuel vapor ignition during fuel filling.[17] Because of this, vehicles using

E85 require a flame arrestor, which is installed into the fuel filling tube. This device will extinguish any flame that might occur.

EFFECT OF E85 ON THE ENVIRONMENT

One of the main motivations for using E85 is its ability to help reduce the impact of vehicle emissions on the environment. E85 provides major reductions in some tailpipe emissions compared to gasoline. E85 is also less toxic than gasoline. Furthermore, the ethanol used in E85 can be derived from renewable resources, thus reducing net greenhouse gas emissions.

VEHICLE TAILPIPE EMISSIONS

Determining the effect of E85 on vehicle emissions is complex, since many factors influence the emissions of vehicles. Further, E85 use is a politically charged issue, effecting the environment, domestic employment, and petroleum imports. Finding reliable emissions data is, therefore, challenging. Actual emissions will vary with engine design and calibration. One of the more recent sources, the U.S. Environmental Protection Agency (EPA), reports potential substantial tailpipe emissions benefits when using E85 relative to conventional gasoline.[18] These benefits are shown in Figure 8.1. This source suggests these benefits for an engine optimized to operate on E85. The EPA also reports that fewer total toxics are produced, and that the hydrocarbon emissions have a lower reactivity. The use of E85 does produce higher ethanol and acetaldehyde emissions than gasoline.

Other sources provide different values, but most sources tend to show substantial reductions in carbon monoxide. For example, the Renewable Fuels Association reports a reduction of 25%.[19] E85 typically results in slightly reduced levels of unburned hydrocarbons. Emissions of nitrous oxides (NOx) are slightly

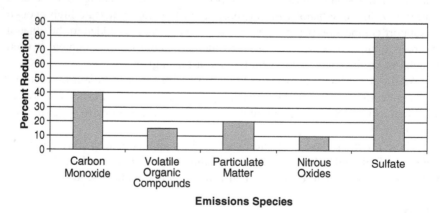

FIGURE 8.1 Estimated emissions reductions for an engine optimized to use E85 compared to those when operating on gasoline. *Source*: Data compiled from U.S. Environmental Agency, EPA420-F-00-035, Mar. 2002.

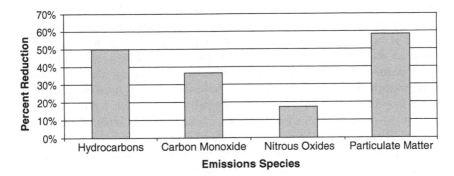

FIGURE 8.2 Reduction in emissions when using E85 compared with E10 for a clean snowmobile. *Source*: Davis, G. and Pilger, C., American Institute of Aeronautics and Astronautics, AIAA-2004-5681, 2004.

reduced with some sources showing slight increases and others showing decreases. Again, much of the data is subject to the test schedule used, and the vehicle and its optimization for E85.

Data from a snowmobile powered by a four-stroke, spark-ignited engine modified to operate using blends up to E85 is shown in Figure 8.2.[20]

If the ethanol used in E85 comes from renewable resources such as corn, E85 can show substantial reductions in greenhouse gas emissions. In 1998, the U.S. DoE Argonne National Laboratory estimated that 1 gallon of E85 reduces greenhouse gas emissions by 16–28% compared to gasoline.[21] Other references suggest higher reductions exceeding 50%.[22]

OTHER ENVIRONMENTAL EMISSIONS

E85 is safer than gasoline to store, transport, and refuel. Ethanol is water soluble and biodegradable. Land and water spills of pure ethanol are usually harmless, dispersing and decomposing quickly. In E85 spills, the gasoline portion of a spill is still a problem; however, the total volume of gasoline in the spill is reduced. This is significant as it is estimated that the amount of oil leaked from vehicles into rivers, lakes, and groundwater is estimated to be six times the annual volume of oil spills.[23]

Finally, since E85 has fewer highly volatile components than gasoline, it produces lower levels of emissions resulting from evaporation. This leads to lower vehicle evaporative emissions during fueling and when the vehicle is not being used. This, of course, is what causes the problems in cold starting a vehicle using pure ethanol, which is why E85 is used in vehicles instead of pure ethanol.

SUSTAINABILITY

Ethanol can be made from biomass resources containing sugar. The main source of ethanol in the United States is corn. Over the years, considerable debate surrounded the energy balance for the production of fuel ethanol from corn. An early study,[24] found that it takes more energy to produce and distribute fuel ethanol than is recovered when combusting the fuel. This led critics to proclaim that corn ethanol is not a fossil energy substitute and its use would actually increase petroleum use. More recent studies have shown that ethanol production is net energy positive, that is burning ethanol liberates more energy than is consumed in its production.[22] One of the most comprehensive studies from an independent source found that 1.34 BTU of energy are delivered for each 1 BTU input.[25] By contrast, gasoline delivers about 0.8 BTU for each 1 BTU input. Thus, E85 use provides about 1.25 BTU for every 1 BTU of energy input.

While the energy balance is important, the type of energy used and produced is perhaps even more important. For example, ethanol is a high-energy density liquid fuel. E85 can be used to displace the use of fossil energy. This approach focuses the energy balance issue, looking at the energy value of the liquid fossil fuels used in the production of ethanol. Using this approach, 1 BTU of fossil energy (typically petroleum) generates between 4.07 and 6.34 BTU of ethanol fuel energy.[26,25] Converting to E85, this means each 1 BTU of fossil fuel generates between 3.46 and 5.39 BTU of E85 fuel. According to a 1998 report from Argonne National Laboratory, each gallon of E85 used reduces petroleum use by 73–75%.[21]

FUTURE TRENDS IN E85

The future for E85 use in vehicles remains uncertain. The use of ethanol has grown dramatically in the United States, rising from about 175,000 gallons in 1980 to a projected 4.4 billion gallons in 2005.[22] Yet the use of E85 is still relatively small. The United States has over 4.1 million E85 capable vehicles on the road. While this sounds impressive, consider that in 1999 the United States had almost 0.8 vehicles per capita, or over 200 million vehicles in use.[27] Thus E85 vehicles represent only about 2–3% of all vehicles. This means that the consumer demand is not yet high enough to provide economic incentive for the fuel-producing companies to produce large quantities of E85 at the pump. Further, most, if not all, of the E85 capable vehicles are FFVs, which means that they operate well on gasoline, reducing the consumer demand for E85.

Still, as more experience is gained using lower blends of ethanol fuels, governments and fleet operators have shown a greater willingness to use E85. For example in 2005, the state of Minnesota legislature voted to require E20 use, up from the current requirement of E10. Furthermore, the U.S. Postal Service Northern District has a large fleet of E85 FFVs, comprising about 13% of the total fleet.

The turning point for E85 use in vehicles appears to hinge on the total market penetration of FFVs and on the total fuel cost. As the market develops in size, owners of FFVs will demand E85 when it is economically advantageous to use. If this happens then, perhaps, Henry Ford was correct when he proclaimed ethanol the fuel of the future and designed the Model T to operate on ethanol or gasoline.

REFERENCES

1. Port, D. The Road Ahead, *Natural Gas Fuels*, May, 2005.
2. Lambrecht, B., Brazil Offers Model for Ethanol Success,. *St. Louis Post-Dispatch*, May 20, 2005.
3. Pulkrabek, W., Thermochemistry and Fuels,. *Engineering Fundamentals of the Internal Combustion Engine*, 2nd Edition, Pearson Prentice Hall, Englewood Cliffs, NJ, 2004, pp. 173–177.
4. *EPAct: Alternative Fuels for Energy Security, Cleaner Air*, U.S. Department of Energy, DOE/GO-102005-2012, Jan., 2005, http://www.eere.energy.gov/vehicle-sandfuels/epact.
5. *Clean Cities Program*, U.S. Department of Energy, Energy Efficiency and Renewable Energy, http://www.eere.energy.gov/cleancities.
6. RFA Tax Guidance on VEETC. The Renewable Fuels Association, Publication # 050, Feb. 1, 2005.
7. Alternative Fuels Data Center, U.S. Department of Energy, Energy Efficiency and Renewable Energy, http://www.eere.energy.gov/afdc/e85toolkit/eth_vehicles.html.
8. Davis, G. et al., The Effect of Multiple Spark Discharge on the Cold-Startability of an E85 Fueled Vehicle, Society of Automotive Engineers, 1999-01-0609, 1999.
9. Stodert, A. et al., Fuel System Development to Improve Cold start Performance of a Flexible Fuel Vehicle, *Society of Automotive Engineers*, 982532, 1998.
10. Shauck, M., and Znain, M, *The Future is Now: Ethanol in Aviation*, Renewable Aviation Fuels Development Center, Baylor University, 1998.
11. Davis, G., *Reduction In Snowmobile Emissions When Using High Blend Alcohol (E-85) Fuel*, Presented at the 2003 International Energy Conversion Engineering Conference, AIAA, AIAA-2003-6025, 2003.
12. Markel, A.J. and Balley, B.K., *Modeling and Cold Start in Alcohol-Fueled Engines*. National Renewable Energy Laboratory Report NREL/TP 640-24180, 1998.
13. *Handbook for Handling, Storing, and Dispensing E85*, U.S. Department of Energy, DOE/GO-1002001-956, Revised Oct. 2002.
14. Sinor, J. and Bailey, B., Current and Potential Future Performance of Ethanol Fuels, *Society of Automotive Engineers*, 930376, 1993.
15. Bolt, J., *A Survey of Alcohol as a Motor Fuel*, SAE Special Publication SAE/PT-80/19, 1980.
16. *2005 Flexible Fueled Vehicles*, U.S. Department of Energy, http://www.fueleconomy.gov.
17. Davis, G. and Heil, E., The Development and Performance of a High Blend Ethanol Fueled Vehicle, *Society of Automotive Engineers*, 200-01-1602, 2000.
18. *Clean Alternative Fuels: Ethanol*. U.S. Environmental Protection Agency, EPA420-F-00-035, Mar. 2002.
19. E85 Can Reduce Ethanol Fueling America's Future Today. Brochure produced by the Renewable Fuels Association, 2001.
20. Davis, G. and Pilger, C., Effect of Biomass-based Fuels and Lubricants on Snowmobiles, *American Institute of Aeronautics and Astronautics*, AIAA-2004-5681, 2004.
21. Corn to Fuel Your Car: A Good Environmental Choice? *TransForum*, Summer 1998.

22. Wang, M., *Fuel-Cycle Energy and Greenhouse Emission Impacts of Fuel Ethanol*, Center for Transportation Research, Argonne National Laboratory, Presentation at EPA Cincinnati, Cincinnati, May 8, 2003.

23. The State of Canada's Environment 1995, *Environment Canada*, 1996.

24. Pimentel, D., Ethanol Fuels: Energy Security, Economics, and the Environment, *J. Agric. Environ. Ethics*, 4, 1991.

25. Shapouri, H. et al., *The Energy Balance of Corn Ethanol: An Update*, U.S. Department of Agriculture, Agricultural Economic Report Number 813, 2002.

26. *Fuel Cycle Evaluations of Biomass Ethanol and Reformulated Gasoline*, Biofuels Systems Division, U.S. Department of Energy, DOEYGO100094.002, Jul. 1994.

27. *Environmental Data Compendium 1999,*. Organization for Economic Cooperation and Development, OECD, Paris, 1999.

Section III

Applications of Alcoholic Fuels

9 Current Status of Direct Methanol Fuel-Cell Technology

Drew C. Dunwoody, Hachull Chung,
Luke Haverhals, and Johna Leddy*
University of Iowa, Department of Chemistry,
Iowa City

CONTENTS

INTRODUCTION

This chapter presents aspects of current developments in direct methanol fuel cell (DMFC) technologies. In particular, the focus is on systems where the fuel stream is a solution of water and methanol fed directly to the fuel cell anode. These systems are the primary focus of the review as cells using reformed methanol as the fuel is a subset of the general class of polymer

---

* Author to whom correspondence should be addressed.

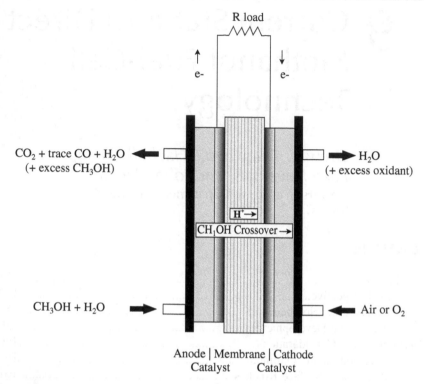

R load

e-   e-

CO_2 + trace CO + H_2O ◀━━━   ━━▶ H_2O
(+ excess CH_3OH)   (+ excess oxidant)

$\boxed{H^+ \rightarrow}$
CH_3OH Crossover →

$CH_3OH + H_2O$ ━━▶   ◀━━ Air or O_2

Anode | Membrane | Cathode
Catalyst   Catalyst

FIGURE 9.1 Diagram of a DMFC. A methanol and water solution is fed to the anode and oxidant is fed to the cathode. The catalyst layers serve as the sites for electrolysis and the membrane serves as the electrical separator and ionic conductor. Methanol crossover from anode to cathode causes parasitic power loss. Reducing the degree of methanol crossover is an active area of DMFC research.

electrolyte membrane (PEM) fuel cells that derive hydrogen from reformate sources such as alcohols and hydrocarbons. A diagram of a DMFC is shown in Figure 9.1. A short, historical, retrospective of relevant technologies and the thermodynamics that govern methanol-based fuel cell systems are presented. A survey of the state of the art, largely in terms of performance, are presented for a variety of fuel cell categories. Efficiency and cost are often the driving force behind development efforts. Trends in DMFC design, electrocatalyst, and membrane development are discussed. Finally, some closing thoughts and discussion of one of the more challenging limitations of extant DMFC technology. The chapter focuses on the application side of the technology, and, given space constraints, computer modeling studies are not covered in depth. Save for historical references, pains are taken to use the most current references available. The majority range from 2003 to the present.

HISTORICAL PERSPECTIVE

The development of modern fuel cells has been driven by the need to generate clean and efficient electrical power for different applications. The first demonstration of a fuel cell was described by William Grove in 1839. Grove was inspired by the observation that electrolysis of water produces hydrogen and oxygen gas. He ran the process in reverse by feeding oxygen to a Pt cathode and hydrogen to a Pt anode, where all electrodes were immersed in a common sulfuric acid bath. Several such cells were connected in series to generate voltage that was measured as shown in Figure 9.2. This "gas voltaic battery" was of little practical value.

Notable work done in the late 19th century by Ludwig Mond and Charles Langer aimed to produce a working fuel cell run on air and industrial coal gas. They were the first to suggest the use of "stacks" of cells with manifolds to deliver fuel and oxidant streams. William White Jaques, who is generally credited with coining the term "fuel cell," replaced Grove's sulfuric acid electrolyte with a phosphoric acid electrolyte. All of these systems fell short of producing practical power plants.

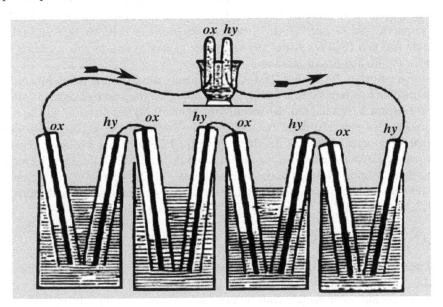

FIGURE 9.2 Diagram of Grove's "gas battery." The cell on top is connected with reversed polarity relative to the four cells on bottom. Water is electrolyzed to hydrogen and oxygen gas in the top cell and fed to the anode and cathode of the bottom cells. Voltage is measured across the top cell. The cell was run several times and performed similarly for each iteration. Image used with permission from *Philosophical Magazine* (http://www.tandf.co.uk).

Francis T. Bacon, direct descendant of the renowned and similarly named philosopher, was first in developing a truly useful fuel cell power plant in 1959. It was a 5-kW system used to power a welding machine. It used nickel electrodes and an alkaline KOH electrolyte. Later that same year, Harry Karl Ihrig demonstrated the first fuel cell-powered vehicle, an Allis Chalmers tractor, powered by 1008 cells split into 112 stacks comprising a 15-kW power source [1,2,3]. Bacon's fuel cell design was the product of more than a quarter century of effort on his part and was the basis for rapid development of fuel cells during the "space race" between the United States and the Soviet Union. Fuel cell development culminated in the use of fuel cells in the Gemini and Apollo missions during the late 1960s and early 1970s [1,2,3]. It is interesting to note that during the earlier Mercury missions and Gemini missions 1 through 4, batteries were used for power. In Gemini 5 and later missions, power was generated by polymer electrolyte fuel cells (PEFCs). In the Apollo missions, PEFCs were replaced by the alkaline fuel cell design because of performance problems caused by oxygen crossover and PEM instability.

The early 1970s to the present can arguably be thought of as the modern era of fuel cell development. A modern fuel cell design generally falls into one of five categories: alkaline fuel cell (AFC), polymer electrolyte fuel cell (PEFC), phosphoric acid fuel cell (PAFC), molten carbonate fuel cell (MCFC), and solid oxide fuel cell (SOFC). All of the categories generate considerable attention in the scientific and patent literature.

Evolution of the modern DMFC is intimately linked to the development of the modern low-temperature PEFC. PEFCs fueled with a pure hydrogen or reformate stream have received the majority of attention in the literature as the top candidate for power systems ranging from the 10^{-3} to 10^5 W needed for power plants. The recent historical development of the PEFC follows a bifurcated path of parallel development of electrocatalysts and PEMs. DMFCs are under increasing consideration as an attractive alternative to PEFCs because of the inherent economic and technological limitations of hydrogen production and storage [1,2,3,4,5].

CATALYST DEVELOPMENT

Pioneering work in DMFC technology was undertaken by Shell, Exxon-Alsthom, Allis Chalmers, and Hitachi during the 1960s and 1970s [1,6]. Research focused on developing noble metal catalysts in liquid acid and alkaline electrolytes [1]. During this period, the mechanistics of methanol oxidation at Pt-based catalysts were studied [1,7,8]. While fundamental understanding of methanol oxidation became more clear, maximum current densities remained low. It was thought the limitation on current density was largely due to inadequate ionic conduction and stability of the PEMs employed in the fuel cells.

MEMBRANE DEVELOPMENT

In the mid 1960s, DuPont introduced the perfluorinated superacid membrane Nafion®. It was considered a major advancement in PEM materials [3]. Earlier, less effective PEM materials included polystyrene-based ionomers and heterogeneous sulfonated divinylbenzene cross-linked polystyrene [2,3]. These early PEMs had poor long-term chemical stability and low proton conductivity. Nafion performance was considerably better than these other materials, however, it was quickly recognized that methanol crossover through Nafion would limit its usefulness as a separator in DMFCs [9]. Crossover diminishes cell efficiency and occurs when fuel that is fed to the anode crosses through the membrane to the cathode and reacts directly with the oxidant. The process also poisons the cathode electrocatalyst with methanol oxidation products.

In the mid-1980s, Nafion membranes became more widely available and solubilized Nafion was introduced to the market. As Nafion became more widely available, PEFC research began in earnest and has since continued. The most notable improvement is a dramatic reduction in catalyst loading at ever-increasing power outputs [6]. Another important discovery was that the stability of the Pt electrocatalyst is greatly enhanced when Nafion is added to the electrocatalyst later. These developments set the stage for a revival of interest in low-temperature DMFCs during the late 1990s [6,10,11,12]. Figure 9.3 demonstrates the increase in DMFC research activity over this period.

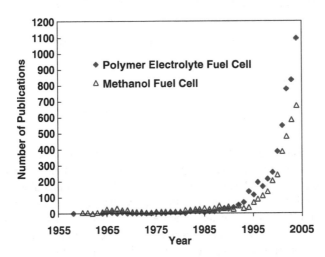

FIGURE 9.3 Plot of hits by year for the topical search "polymer electrolyte fuel cell" (♦) and "methanol fuel cell" (▲) using the SciFinder Scholar 2004 search engine. Searches conducted in May of 2005.

As will be seen in later sections of this chapter, current efforts in DMFC research include minimizing methanol crossover through the separator of DMFCs while maintaining high proton conductivity, developing methanol tolerant oxygen reduction catalysts, and identifying more cost-effective methanol oxidation catalysts [13].

THERMODYNAMIC CONSIDERATIONS

The following is a brief treatment of the thermodynamics governing the methanol oxidation reaction of a DMFC. Also, the impact of surface kinetics on the practical efficiency of the cell are presented. Some intriguing reports suggesting a new general direction for CO-tolerant catalyst development are cited [14,15,16].

THERMODYNAMIC OPTIMUM

When an organic fuel is used, essentially as a hydrogen source in a fuel cell, the expectation is that the fuel will be completely oxidized to carbon dioxide. For methanol, this is summarized thermodynamically [17] in terms of the reduction potentials as

$$CO_2 + 6H^+ + 6e \rightleftharpoons CH_3OH_{(aq)} + H_2O \qquad E^0 = 0.016V \qquad (9.1)$$

While methanol is oxidized at the anode, oxygen is reduced at the cathode:

$$O_2 + 4H^+ + 4e \rightleftharpoons 2H_2O \qquad E^0 = 1.229V \qquad (9.2)$$

The net cell reaction is

$$CH_3OH_{(aq)} + 1.5O_2 \rightleftharpoons 2H_2O + CO_2 \qquad (9.3)$$

where the standard cell potential (electromotive force, emf) is E^0_{cell} = 1.229 − 0.016 = 1.213V. For a six-electron process (n = 6), the standard free energy is $\Delta G^0 = -nFE^0_{cell}$ = −702.2 kJ/mol for methanol. With a molecular mass, M, of 0.03204 kg/mol, the theoretical specific energy for methanol is $W = -\Delta G^0/(M \times 3600s/hr)$ = 6.088kWh/kg; because the density of methanol is 0.7914 kg/l, this corresponds to an energy density of 4.818kWh/l. The standard enthalpy [17], ΔH^0 = −726kJ/mol, is similar to ΔG^0, consistent with a small entropy term.

Formally, the complete oxidation of methanol can be viewed thermodynamically as a series of two-electron/two-proton oxidation steps. The reduction potentials for this sequence in acid are given as [18]:

$$HCHO_{(aq)} + 2H^+ + 2e \rightleftharpoons CH_3OH_{(aq)} \qquad E^0 = 0.232V \qquad (9.4)$$

$$\text{HCOOH}_{(aq)} + 2\text{H}^+ + 2e \rightleftharpoons \text{HCHO}_{(aq)} + \text{H}_2\text{O} \qquad E^0 = 0.034V \qquad (9.5)$$

$$\text{H}_2\text{CO}_{3(aq)} + 2\text{H}^+ + 2e \rightleftharpoons \text{HCOOH}_{(aq)} + \text{H}_2\text{O} \qquad E^0 = -0.166V \qquad (9.6)$$

where the oxidation of formic acid is reported to carbonic acid, consistent with the solubility of carbon dioxide and its equilibrium with carbonate [19].

$$\text{CO}_{2(g)} \rightleftharpoons \text{CO}_{2(aq)} \qquad K_{\text{CO}_2} = 0.034 \qquad (9.7)$$

$$\text{CO}_{2(aq)} + \text{H}_2\text{O} \rightleftharpoons \text{HCO}_3^- + \text{H}_{(aq)}^+ \qquad pK = 6.36 \qquad (9.8)$$

The acidity constants for carbonic acid are $pK_{a1} = 6.352$ and $pK_{a2} = 10.329$; for formic acid, $pK_a = 3.745$. Note that these reaction steps embed information about the complexities of the solution chemistry in the fuel cell as reaction products build and local pH changes. Note also, that reactions are reported in acid because practical DMFCs are usually run under acidic conditions. Under basic conditions, the formation of insoluble carbonates dramatically complicates the design of plant and limits applicability as electrolytes must be replaced as carbonate levels build.

For species in solution, the standard potentials (reactions 4 to 6) are such that thermodynamically, the oxidation of methanol proceeds cleanly and sequentially from alcohol to aldehyde to acid to CO_2/carbonic acid with approximately 200 mV separating each successive two proton/two electron transfer. The specific energy and energy density of methanol are high. Thus, thermodynamically, the expectation is that methanol is an excellent fuel for a direct reformation fuel cell. However, the thermodynamics do not capture the complexity of the surface reactions that dictate the fate of methanol in a direct reformation fuel cell.

REALITIES OF SURFACE KINETICS

The kinetic limitations of DMFCs have been well reviewed in detail from several different perspectives in recent years [17,20,21]; an early and thorough review is provided by Parsons and VanderNoot [22]. For effective utilization of methanol as a fuel, the catalyst must provide a good surface for adsorption of methanol and its sequential breakdown to carbon dioxide/carbonate through loss of paired protons and electrons. Under acidic conditions, this has largely restricted practical catalysts to platinum and its alloys and bimetallics. Methanol will adsorb to platinum and platinum serves as an excellent electron transfer catalyst. The difficulty is that platinum passivates as carbon monoxide by-product accumulates and adsorbs to the platinum surface. To oxidize carbon monoxide to carbon dioxide/carbonic acid, oxygenated species such as water must adsorb to the catalyst surface. Because platinum is not strongly

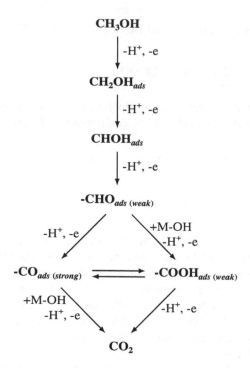

SCHEME 9.1 Reaction pathways for methanol oxidation.

hydrophilic, platinum bimetallics and alloys formed with more hydrophilic metals such as ruthenium are typically used to facilitate CO oxidation.

Consider the mechanistic constraints for oxidation of methanol. As in equation 1, the complete oxidation of methanol to carbon dioxide proceeds by a six-proton, six-electron process. The mechanism presented in Scheme 9.1 outlines the basic route by which methanol is fully oxidized. The loss of paired protons and electrons is noted for each step. To account for all six electrons, recognize that the adsorption of water to the catalyst surface also generates an electron and proton. For a catalyst metal site, M,

$$M + H_2O \rightleftharpoons M - OH + H^+ + e \tag{9.9}$$

Following the notation from Ref. [21], methanol first adsorbs to liberate one electron and one proton.

$$CH_3OH + Pt \rightleftharpoons Pt - CH_2OH_{ads} + H^+ + e \tag{9.10}$$

This is followed by two steps to form the formyl intermediate, –CHO.

$$Pt - CH_2OH_{ads} + Pt \rightleftharpoons Pt_2CHOH + H^+ + e \qquad (9.11)$$

$$Pt_2CHOH \rightleftharpoons Pt + Pt - CHO + H^+ + e \qquad (9.12)$$

On clean platinum surfaces, these oxidations proceed smoothly to provide two electrons and two protons. Consider Scheme 9.1. The weakly adsorbed –CHO is a point at which the oxidation mechanism breaks into two paths. One path yields adsorbed CO and the other adsorbed COOH. Adsorbed COOH is generated by reaction of –CHO and an adjacent M – OH to yield one proton and one electron and form weakly adsorbed –COOH. Adsorbed CO is generated by the direct oxidation of –CHO by one proton and one electron to form strongly adsorbed CO. Basic kinetic arguments would favor the strongly adsorbed CO over the weakly adsorbed –COOH because first, the oxidation of –CHO to –CO is direct and does not require an adjacent second species, M – OH, and second, because –CO is strongly bound and –COOH is weakly bound.

It should be pointed out that there is an alternative branch point in the oxidation process in which adsorbed –CHOH undergoes a one-electron and a one-proton oxidation to form adsorbed –COH.

$$Pt_2CHOH + Pt \rightleftharpoons Pt_3COH + H^+ + e \qquad (9.13)$$

The adsorbed –COH can then either undergo one-proton/one-electron oxidation to adsorbed –CO or react with an adjacent M – OH to form HCOOH in solution. Neither process leads to the efficient oxidation to carbon dioxide/carbonic acid.

To the extent the platinum surface is passivated by CO, the reaction is terminated. Thus, the design of a system for the efficient and complete oxidation of methanol can be approached in two ways.

The first approach is to circumvent the formation of adsorbed CO by favoring the formation of –COOH. Experimentally, this is done by enhancing the probability that –CHO is adjacent to an oxygen source, M – OH, by using bimetallics and alloys of platinum where M is more hydrophilic than platinum. There are questions of stability and cost associated with these catalysts although they have been shown to enhance conversion efficiency. But, based on the relative strengths of the adsorbates –CO and –COOH and the need for an additional catalyst site (M – OH), this approach poses some challenges.

The second approach is to consider why –CO is so difficult to oxidize; that is, why does CO adsorb so strongly. Thermodynamically, the oxidation of CO to CO_2 in solution occurs at low potential [18].

$$CO_2 + 2H^+ + 2e \rightleftharpoons CO + H_2O\,(1)\ in\ E^0 = -0.106V \qquad (9.14)$$

But, the oxidation of CO on platinum in acidic solution occurs 600 to 700 mV positive of this value; Pt-Ru alloys are shown to oxidize CO at 200 to 300 mV lower overpotential than Pt [23]. The oxidation of adsorbed CO is strongly disfavored. There are two ways to think about overcoming this large overpotential. One is to design better catalysts. One common approach has been through the bifunctional mechanism where the bimetallic catalyst is designed to place $Pt -$ CO adjacent to an oxygen source through M – OH. The other approach would rely on a paradigm shift in how the oxidation of –CO is viewed at a more fundamental level; better understanding could lead to better catalysts [14,15,16].

The above discussion is provided in a very general manner. Many factors significantly impact the catalytic efficiency of the conversion of methanol to carbon dioxide/carbonic acid. This includes surface structure, catalyst size, and catalyst crystal face as well as the history of the cell, the current coverage of CO, the pH, and the time since the start of the cell.

PERFORMANCE TARGETS AND EFFICIENCIES

Discussions of performance targets and efficiencies for DMFCs are complicated due to the wide-ranging conditions, fuel and oxidant sources, and intended applications for DMFCs. In this section, a survey of performance data and examples of target system requirements listed by government agencies are used to give a sense of the state of the art. Also, targets set by researchers in the literature are discussed.

In 2002, Jörissen et al. suggested DMFC performance targets to compete in terms of efficiency with reformate fed PEFCs [24]. The target they set for a DMFC is a power density of 250 mW cm^{-2} at a cell voltage of 500 mV and that furthermore, parasitic power loss due to methanol crossover should be no more than 50 mA cm^{-2} at a power density of 250 mW cm^{-2}.

In a 1999 review of advanced electrode materials for use in DMFCs, Lamy and Léger discussed the suitability of a number of energy systems in relation to DMFCs for use in automobiles [17]. Secondary batteries (e.g., Li-ion) are limited by recharge time and power density (100–150 Wh kg^{-1} at maximum). PEFCs are attractive with specific power densities on the order of 1000 W kg^{-1} and specific energy density >500 Wh. Energy density of pure H$_2$ is 33 kWh kg^{-1} but storage concerns make it less attractive and less efficient. Performance characteristics of DMFCs circa 1999 is 200 mA cm^{-2} at 0.5 V, or 100 mW cm^{-2} with electrocatalyst loadings under 1 mg cm^{-2}.

Performance targets for a complete DMFC power system were posted in the Spring of 2005 by the U.S. Army Operational Test Command (OTC). The specifications are target requirements for a ruggedized DMFC power plant for use in the field on armored and other military vehicles [25]. The specifications outline threshold requirements and objective targets for the power system. A summary of the requirements are listed in Table 9.1. In an effort to meet the objectives listed in Table 9.1, a 300-W prototype DMFC power plant was developed by T. Valdez and his team at the Jet Propulsion Laboratory [26]. The demonstration

TABLE 9.1
U.S. Army OTC Threshold and Objective DMFC Power System Targets

Parameter	Threshold	Objective
Power output	200 W	300 W
Continuous operation duration	70 hours	100 hours
Dimensions	3.5 to 4.0 ft³	2.5 ft³
System weight (with fuel)	110 lbs	95 lbs
System weight (without fuel)	60 lbs	45 lbs
Voltage out	24 VDC	12/24 VDC
Start-up time (at 0°C)	10 minutes	5 minutes
Operating life	4000 hours	7000 hours
Efficiency (system output/stackoutput)	60%	75%
Shelf life	3 years	5 years
Noise	Not audible beyond 25 ft	Not audible beyond 8 ft
Dust concentration tolerance	20 × zero visibility 5 gm m⁻² ACS coarse 30 μm dust	Same
Thermal signature	Ambient	Ambient
Outside operating temperature	0 to 50°C	0 to 70°C
Storage temperature	0 to 40°C	−10 to 50°C

power plant was designed for 100 hours of continuous operation and used 80 cells with active areas of 80 cm². The electrocatalyst was PtRu at the anode and cathode. The plant generated 370 W during bench testing and had a start-up time of 18 minutes. The plant was operated continuously for 8 hours, generating a lower than expected power of 50 W. The continuous operation test was ended due to water accumulation in the stack exhaust manifold.

Subsequent to testing of the prototype power plant, the stack was torn down and components evaluated. The wettability of the cathodes of the MEAs had increased and evidence of the ruthenium migration was observed. These observations were the impetus to study of the long-term stability of DMFC MEAs. The team at JPL individually ran four MEAs on a single-cell test stand for 250 hours. All of the MEAs showed irreversible voltage decay ranging from 0.2 to 0.6 mV hr⁻¹ at a current density of 100 mA cm⁻² that resulted in an average decline in power of 20%. However, unlike when the MEAs were run as components of the stack in the prototype power plant, the individually run MEAs showed no evidence of electrocatalyst migration. The important issue of electrocatalyst migration will be addressed again in the final section of this chapter.

According to Knights et al. at Ballard Power Systems, fuel cell power plants used in automobile, bus, and stationary applications require operational lifetimes on the order of 4000, 20,000, and 40,000 hours, respectively [27]. The degradation rate of the power supply is set by the beginning-of-life (BOL) and end-of-life (EOL) performances; a degradation rate on the order of 10 to 25 μ V hr⁻¹ is common for DMFCs. The group studied the strategy of load cycling in DMFCs to reduce performance degradation caused by water build-up at the cathode with time.

Ball Aerospace is developing a personal DMFC power system to meet the needs of the U.S. foot soldier [28]. It was developed under the Defense Advanced Research Projects Agency (DARPA) Palm Power program and produces average power of 20 W at 12 V and has a 30-W peak power. The unit operates for 50 hours on the fuel provided by one fuel cartridge, and is ten times lighter than the equivalent battery power plant; weighing in at three pounds with full fuel complement.

Yi et al. characterize the changes in MEA morphology of a single-cell DMFC run for a little longer than three days [29]. Long-term stability of the cell and electrocatalyst are important questions. The cell was run at 100 mA cm^{-2} and suffered from irrecoverable performance degradation, degrading at the rate of 1.0 to 1.5 mV hr^{-1}. Following the run, Yi and his group found signs of delamination between the layers of the MEA and that both of the carbon-supported electrocatalysts, PtRu/C on the anode and Pt/C on the cathode, had undergone a particle size redistribution resulting in larger particle sizes on average. The redistribution for the PtRu electrocatalyst was more pronounced than for Pt and more severe in the anode.

An assessment of the state of the art in DMFC performance can be made from relevant data from references in this chapter; data are listed in Tables 9.2 and 9.3. Where possible, the data listed from a particular reference includes data for the "best" test cell and the associated control cell. The best test cell is considered the one with highest maximum power density. The control cell is usually of a typical Nafion MEA construction consisting of carbon-supported PtRu on the anode, carbon-supported Pt on the cathode and a Nafion 115 membrane as the separator. Efforts have been made to include operating conditions and loadings. Where an entry is listed as "n/a" the value for that parameter is not available. That is, the reference does not explicitly state the value of that parameter.

GENERAL OPTIMIZATION

The following section cites selected studies aimed at optimizing the performance of DMFCs through careful variation of design, materials, and operating conditions. An excellent study of a wide range of experimental conditions is presented first, then issues of cathode flooding, electrolyte/electrode contact, parasitic power loss associated with fuel pumping, electrode design, and CO$_2$ bubble formation are considered.

A systematic study by Liu and Ge varied operational parameters such as cell temperature, methanol concentration, anode flow rate, air flow rate, and cathode humidification, and showed that changing any one of the parameters has a pronounced effect on the performance of the DMFC [30]. However, varying cathode humidification has negligible impact on DMFC performance. The range of parameters evaluated are listed in Table 9.4. In general, higher cell temperatures lead to better DMFC performance; however, other processes that diminish cell performance as temperature increases such as methanol crossover and water transfer from cathode to anode set a limit on optimum performance. The study found the optimal methanol solution to have a concentration between 1 and 2 M. This is in general agreement with other studies that found the optimum concentration to be 2.0 M [5] and 2.5 M [31] for DMFCs run under similar operating conditions. Methanol crossover from the anode to the cathode can be minimized by increasing either the cathode air flow rate or oxygen partial pressure. The work suggests that the cathode structure and operating conditions play a major role in DMFC performance. The reference by Liu and Ge contains a large amount of data, both plotted and tabulated, and is a useful resource for making comparisons of DMFC performance over a range of conditions.

For DMFCs power plants, performance degradation occurs when water builds up at the cathode. Knights et al. describe a load-cycling strategy to reduce cathode flooding [27]. By removing the load of the DMFC for 30 seconds of every 30 minutes of operation, the rate for performance degradation is shown to be 13 μV hr^{-1} over 2000 hr of failure-free operation, which is in the low range of the typical performance degradation rate of 10 to 25 μV hr^{-1}.

The use of solid PEMs such as Nafion prevents electrolytes from fully enveloping the electrode as liquid electrolyte does. This limits the reaction area to points of direct contact between membrane and electrode. To increase the reactive area, Nafion suspension is often compounded directly into the catalyst layer. Sudoh et al. use the spray method to optimize the electrochemical characteristics of the catalyst layer [37]. The spray method consists of introducing a catalyst to the electrode surface and then spraying Nafion over the catalyst. Three Nafion loadings are considered: 0.5, 1.0, and 3.5 mg cm^{-2}. The resulting electrodes are incorporated into DMFCs and the performance measured. The cell made with catalyst layers having a Nafion loading of 1.0 mg cm^{-2} performed the best producing 258 mA cm^{-2} at 0.4 V. The cell made from electrodes with 0.5 mg cm^{-2} Nafion loading generated roughly half the current and the cell with the highest Nafion loading was resistive and performed poorly. The performance of the spray-coated, in-house DMFC is similar to that of commercially available ELAT® electrode with similar catalyst loading (0.5 mg cm^{-2}). ELAT is the trademarked name of gas diffusion electrode material distributed by E-Tek, Inc. It is frequently used in PEFCs and DMFCs.

Zhang and Wang present a piezoelectric micropump design for delivering fuel to a miniaturized DMFC power source [39]. For low current densities (<100 mA cm^{-2}) methanol concentrations between 0.5 M and 2.0 M do not significantly impact the power generated. The authors suggest their DMFC running at 40°C will have a maximum current density, or J_{max} of 120 mA cm^{-2} at 0.35 V for a 1-W system where a 25 cm^2 cell will be required. The estimated power consumption of their piezoelectric pump operating at 100 Hz, pushing 1 ml min^{-1} over the face of the cell is on the order of 70 mW, or 7% of the power produced by the cell, which compares well with other literature examples.

Typical DMFC anode structures consist of strata of a supported/unsupported catalyst bonded with a Nafion suspension over Teflon-bonded carbon black

168   Alcoholic Fuels

TABLE 9.2

Survey of DMFC Performance

Separator	Ano. Catalyst	Loading ($\frac{mg}{cm^2}$)	Anode Flow ($\frac{ml}{min}$)	Cath. Catalyst	Loading ($\frac{mg}{cm^2}$)	Cath. Press. MPa	Cath. Flow ($\frac{ml}{min}$)	Cell T (°C)	[MeOH] M	Max. Power ($\frac{mW}{cm^2}$)	Potential (mV)	Ref.
Naf. 117[a]	Pt-Ru	1.0	1.0	Pt	1.0	0.1	75 air	23	1.0	2.8	180	[32]
Naf. 117	Pt-Ru	1.0	1.0	Pt	1.0	0.1	75 air	23	1.0	0.9	160	[33]
Naf. 117[b]	Pt-Ru	2.0	2.0	Pt	2.0	0.1	500 O_2	40	2.0	36	380	[33]
Naf. 117	Pt-Ru	2.0	2.0	Pt	2.0	0.1	500 O_2	40	2.0	26	100	[34]
Naf. 117[c]	Pt, Ru/C	2.0	n/a	Pt	2.0	0.1	n/a O_2	95	2.0	82	390	[34]
Naf. 117	Pt, Ru/C	2.0	n/a	Pt	2.0	0.1	n/a O_2	95	2.0	32	360	[24]
Naf. 117	PtRu	5.0	4.0	Pt	6.0	0.4	1500 air	110	1.0	120*	500	[24]
Naf. 105	PtRu	5.0	4.0	Pt	6.0	0.4	1500 air	110	1.0	270*	500	
Polyaryl[d]	PtRu	5.0	4.0	Pt	6.0	0.4	1500 air	110	1.0	230*	500	[35]
Naf. 115[e]	PtRu	n/a	n/a	Pt	n/a	n/a	300 O_2	65	2.0	49	350	
Naf. 115	PtRu	n/a	n/a	Pt	n/a	n/a	300 O_2	65	2.0	40	350	[36]
Naf. 115	PtRu/NT[f]	0.4	2.0	Pt/C	0.4	0.1	80 O_2	70	2.0	70	200	
Naf. 115	PtRu/C	0.4	2.0	Pt/C	0.4	0.1	80 O_2	70	2.0	50	200	
Naf. 115	PtRu/NT[f]	0.4	2.0	Pt/C	0.4	0.3	80 O_2	90	2.0	140	250	
Naf. 115	PtRu/C	0.4	2.0	Pt/C	0.4	0.3	80 O_2	90	2.0	125	250	
Naf. 117	PtPu/C	2.0	3.0	Pt/C[g]	2.0	0.2	350 O_2	90	2.0	103*	400	[37]
sPEEK	N/a	n/a	n/a	n/a	n/a	0.3	n/a	120	n/a	140	n/a	[38]
Polymer[h]	N/a	n/a	n/a	n/a	n/a	n/a	n/a	110	n/a	300	n/a	
Polymer[i]	N/a	n/a	n/a	n/a	n/a	n/a	n/a	60	n/a	50	n/a	
Naf. 117	PtRu	3.0	6.0	Pt	3.0	n/a	600 air	70	2.0	66	270	[30]†

Naf. 117	PtRu	3.0	6.0	Pt	3.0	n/a	1200 air	70	2.0	83	330	
Naf. 117	PtRu	3.0	6.0	Pt	3.0	n/a	600 O_2	70	2.0	132	243	[39]
Naf. 117[j]	PtRu	4.0	0.5	Pt	4.0	0.1	20 O_2	40	1.0	40	350	
Naf. 112	PtRu/C	3.0	9.0	Pt/C	3.0	0.3	105 O_2	120	2.5	127*	550	
Naf. 117	PtRu	1.0	n/a	Pt	4.0	0.15	50 air	80	0.5	50*	500	

[a] Membrane has 1-μm sputtered Pd-Ag film

[b] Pd impregnated Nafion; 0.0214 g Pd/cm^3 of Nafion

[c] Pd sputtered membrane

[d] Polyaryl blend of PEK, PBI, and bPSU

[e] Nafion has 14% mass gain from polystyrene

[f] Multiwall carbon nanotube support with Fe- and Ni-contaminated catalyst

[g] Cathode impregnated with 1.0 mg cm^{-1} Nafion ionomer

[h] Sulfonated poly(4-phenoxybenzoyl 1,4-phenylene)

[i] Sulfonated poly[bis(3-methylphenoxy) phosphazene]

[j] Power calculated from model based on Nafion 117 DMFC data [40] minus power needed to drive piezoelectric pump

* Not necessarily maximum power. † Reference has performance data over wide range of conditions

TABLE 9.3
Survey of DMFC Performance

Separator	Ano. Catalyst	Loading ($\frac{mg}{cm^2}$)	Anode Flow ($\frac{mL}{min}$)	Cath. Catalyst	Loading ($\frac{mg}{cm^2}$)	Cath. Press. MPa	Cath. Flow ($\frac{mL}{min}$)	Cell T (°C)	[MeOH] M	Max. Power ($\frac{mW}{cm^2}$)	Potential (mV)	Ref.
Naf. 117	PtRu/C	1.0	12.0	Pt/C	1.0	0.1	1.0 air	90	1.0	64	300	[41]
Naf. 117	PtRu[k]	1.0	12.0	Pt/C	1.0	0.1	1.0 air	90	1.0	58	250	
Polymer[l]	Pt-Ru	4.0	25.0	Pt	4.0	n/a	3000 O$_2$	80	1.0	316	722	[42]
Naf. 115	Pt-Ru	4.0	25.0	Pt	4.0	n/a	3000 O$_2$	80	1.0	309	696	
Naf. 1135[m,n]	Pt-Ru	n/a	n/a	Pt	n/a	0.08	n/a air	70	0.3	29	375	[43]
Naf. 1135[m,o]	Pt-Ru	n/a	n/a	Pt	n/a	0.08	n/a air	70	0.3	51	543	
Nafion[p]	Pt-Ru/C	2.0	n/a	Pt/C	2.0	0.25	n/a O$_2$	145	2.0	400	900	[44]
Nafion[p]	Pt-Ru/C	2.0	n/a	Pt/C	2.0	0.05	n/a O$_2$	145	2.0	200	485	
Nafion	Pt-Ru/C	2.0	n/a	Pt/C	2.0	0.25	n/a O$_2$	145	2.0	350	318	[45]
Naf. 115[q]	Pt-Ru/C	1.3	1.0	Pt/C	1.0	0.2	160 O$_2$	75	1.0	46	380	[29]
Naf. 115[r]	Pt-Ru/C	1.3	1.0	Pt/C	1.0	0.2	160 O$_2$	75	1.0	66	560	[46]
Naf. 115[s]	PtRu	3.0	5.0	Pt/C	3.0	0.1	250 air	80	2.0	70	507	[46]
Naf. 115	PtRu	3.0	5.0	Pt/C	3.0	0.1	250 air	80	2.0	32	320	[47]
Naf. 115	PtSn/C	1.3	1.0	Pt/C	1.0	0.2	n/a O$_2$	90	1.0	17	150	
Naf. 115	PtRu/C	1.3	1.0	Pt/C	1.0	0.2	n/a O$_2$	90	1.0	55	210	
Naf. 115	Pt/C	2.0	1.0	Pt/C	1.0	0.2	n/a O$_2$	90	1.0	136	335	
Naf. 115	PtRu/C	1.0	2.5	Pt/TiO$_2$/C[t]	1.0[v]	0.27	2.5 O$_2$	70	3.0	21	229	[48]
Naf. 115	PtRu/C	1.0	2.5	Pt/TiO$_2$/C[u]	1.0[v]	0.27	2.5 O$_2$	70	3.0	15	165	
Naf. 115	PtRu/C	1.0	2.5	Pt/C	1.0	0.27	2.5 O$_2$	70	3.0	12	177	
Naf. 115	PtRu/C	2.0	1.0	PdPt/C[w]	1.0	0.2	n/a O$_2$	75	1.0	95	320	[49]

Naf. 115	PtRu/C	2.0	Pt/C	1.0	0.2	n/a O$_2$	75	1.0	68	322	
Naf. 115	PtRu/C	2.0	PtFe/MWNT	1.0[x]	0.2	1.0 O$_2$	90	1.0	65	272	[50]
Naf. 115	PtRu/C	2.0	Pt/MWNT	1.0		1.0 O$_2$	90	1.0	54	241	
H$_2$SO$_4$	Pt	n/a	Pt	n/a	n/a	n/a O$_2$	60	n/a	40	n/a	[51]
KOH	PtPd	n/a	Ag	n/a	n/a	n/a O$_2$	60	n/a	25	n/a	

[k] Anode made from Ti mesh with electrodeposited catalyst layer

[l] Membrane is 5-μm thick copolymer of TFE and ethylene

[m] MEAs made of half cells placed back to back so Nafion thickness between electrodes is 7 mil and potential listed for cell is iR corrected

[n] MEA run as component of 22-cell DMFC stack for 6 months

[o] Same cell as listed with "n" superscript prior to use in DMFC stack

[p] 100-μm thick PEMs made from recast Nafion and 3 wt % SiO$_2$ – PWA filler

[q] Single-cell DMFC after 75-hour lifetime test

[r] Same cell as listed with superscript "q" superscript prior to 75-hour lifetime test

[s] 1 wt % CeO$_2$ doped cathode

[t] Catalyst heat treated at 500°C

[u] No heat treatment

[v] Listed value is Pt loading; Pt to Ti ratio of 1:1

[w] Ratio Pd to Pt is 3:1

[x] Ratio Fe to Pt is 1:1; Pt loading 1.0 mg cm^{-2}

TABLE 9.4
Operating Conditions and Range over
Which DMFC Performance is Evaluated

Parameter	Range
Cell Temperature	30–80°C
Methanol Concentration	1–6 Mol L^{-1}
Cathode Humidification Temperature	40–90°C
Anode Flow Rate	0.5–10 ml min^{-1}
Air Flow Rate	100–2000 sccm

diffusion layer over carbon cloth or paper diffusion layer. This structure is an ineffectual design for the transport and release of CO_2 gas produced by methanol oxidation and limits methanol transport to the anode. To remedy this problem, Scott et al. directly deposit PtRu catalyst onto a titanium mesh by electrodeposition and subsequent thermal decomposition and use the coated mesh as the anode [41]. Scanning electron microscopy (SEM), energy dispersive X-ray (EDX) and X-ray diffraction analysis (XRD) are used to characterize the electrodes. Anodes are tested under galvanostatic control as well as in DMFCs. Galvanostatic testing shows the mesh and conventional anodes have similar electrochemical performance. This is somewhat unexpected given the very different morphologies of the strata and electrodeposits of electrode types. Tests using the anodes under working DMFC conditions mirror the performance of the anode tests. Catalysts loadings are in the range of 0.8–1.0 mg cm^{-2}.

An often overlooked limitation in DMFC performance is CO_2 bubble formation in the anode. Kulikovsky developed a simple DMFC model to determine how anode channel bubble formation impacts cell performance [52]. Under conditions simulating typical operating conditions, the model suggests that moderate to severe bubble formation decreases the mean methanol concentration as it passes through the anode channel, limiting the current that can be drawn for the cell. Under severe bubbling conditions, the limiting current that can be drawn from the cell is diminished by as much as a factor of four. The author speculates that faster flow rates may help offset some performance losses due to bubble formation and offers some calculations to support his speculation, but he also cautions that kinetics of bubble formation are outside the scope of the model.

ELECTROCATALYSTS AND SUPPORTS

DMFC electrocatalysts set the catalytic efficiency that dictates the DMFC performance and establish a large component of DMFC cost. Here, we discuss some recent developments in DMFC electrocatalysts and the materials used to support them.

Gonzalez et al. evaluated Pt and PtRu supported on single-wall carbon nanotubes (SWNTs) and multiwall carbon nanotubes (MWNTs) as electrocatalysts [36]. The materials are integrated into electrodes and hot pressing with Nafion 115 to form MEAs. Half-cell experiments show the PtRu/C electrocatalysts have an earlier onset of methanol oxidation (i.e., lower potential) than the Pt/C counterparts. An MEA made with PtRu/MWNT showed the highest activity. When the electrocatalysts are evaluated in DMFCs, it was found the PtRu/C electrocatalysts performed better than their Pt/C counterparts. The sequence of electroactivity being PtRu/MWNT > PtRu/C > PtRu/SWNT. Maximum power densities greater than 100 mW cm^{-2} are obtained with PtRu catalyst loadings of 0.4 mg cm^{-2}.

A modified alcohol reduction method was used by Hwang et al. to produced nanosized PtRu/C electrocatalysts [53]. Various electrocatalyst preparations are compared to a commercial PtRu/C electrocatalyst, E-Tek 40 for morphology and effectiveness as an electrocatalyst. Transmission electron microscopy (TEM) show in-house preparations similarly well-dispersed as the commercial electrocatalyst. Metal particle size can be tailored by selection of the concentrations of preparation components. Activity of the electrocatalysts are compared under various methanol conditions in half-cell measurements made in a three-electrode cell. The results are mixed. Under realistic operating conditions (e.g., 40°C, 0.4 V, and [MeOH] ~ 15%) one of the in-house electrocatalysts outperforms the commercial catalyst. However, at higher potentials, the commercial catalyst performs better. This same trend holds true at [MeOH] = 35% and 50%, where the in-house catalyst performs better than the commercial at 0.4 V, but the commercial performs better at higher potentials. AC impedance data for this in-house catalyst suggest it has a lower resistance at all potentials and for all concentrations of methanol considered in the study.

An alternative to the costly Ru often employed in Pt-based bimetallic electrocatalysts of DMFCs may be Sn. The impact of introducing Sn to the Pt/C anode electrocatalyst of direct alcohol fuel cells (i.e., methanol and ethanol) was evaluated by Zhou et al. [47]. Cyclic voltammograms recorded vs. SCE showed the Sn-bearing electrocatalyst, PtSn/C, had a more favorable onset of oxidation potential (20 mV) than Pt/C (250 mV) and PtRu/C (110 mV), however, the peak oxidation potential of PtSn/C was intermediate (640 mV) to Pt/C (700) and PtRu/C (500 mV). For further comparison, three single-cell DMFCs were prepared. The maximum power densities of the three cell tracked with the peak oxidation potentials of the cyclic voltammetric experiments: PtRu/C (136 mW cm^{-2}) > PtSn/C (55 mW cm^{-2}) > Pt/C (17 mW cm^{-2}). The work points to the possibility that adding relatively inexpensive Sn to the Pt/C anode may significantly improve DMFC catalysts. The effectiveness of this approach is difficult to assess as the performance of the Pt/C blank is so poor, however, the performance trends warrant further research.

One strategy to limit the effects of methanol crossover in DMFCs is to develop cathode electrocatalysts active toward the oxygen reduction reaction (ORR) but inactive toward the methanol oxidation reaction (MOR). Pt is active toward both reactions whereas Pd is active only toward ORR. In a short communication, Sun et al. use voltammetry and chronoamperometry to demonstrate a Pd:Pt alloy in the ratio of 3:1 supported on carbon is effective toward the ORR and ineffective toward the MOR [49]. The group goes on to compare the performance of DMFCs impregnated with either the Pd_3Pt_1/C electrocatalyst or Pt/C control in the cathode. The cell using the PdPt alloy had better overall performance with a maximum power density, ~40% higher than that of the Pt/C control. The authors think that once the ratio of Pd to Pt rises above a certain point, the active sites of the Pt become isolated and overall activity of the alloy toward MOR drops precipitously.

The incorporation of the oxygen storage material CeO_2 into the carbon supported Pt electrocatalysts of the cathode was found to enhance the performance of DMFCs when run on air [46]. The ceria compound is known to act as an oxygen storage buffer and helps to maintain the local oxygen pressure, but the effect only appears to be beneficial when using air. Yu et. al. found that when run on pure oxygen, the presence of ceria oxide in the cathode diminished the maximum power of a DMFC by ~20%. The most pronounced enhancements on performance are when operating the cell at low air-flow rates (250 sccm), but the effect becomes minimal at higher flow rates (1250 sccm). Impedance spectroscopy was used to determine the polarization resistance of the ceria-doped and nondoped Pt/C cathodes under air and O_2. The resistance of the ceria-doped cathode was lower than the control when under an air atmosphere, but the ceria-doped cathode had a higher resistance under O_2. An optimized cathode composition for use in air was found to be 1 wt % CeO_2 and 40 wt % Pt/C.

The performance of PEFCs run on H_2 and O_2 has been shown to improve upon the incorporation of TiO_2 into the carbon-supported Pt electrocatalyst layer [54]. Manthiram and Xiong tested the same modification of the DMFC electrocatalyst layer [48]. A number of deposition methods were tried as well as a series of heat treatments under a reducing atmosphere. The treatments result in an array of different sized particles (3.8 to 25.4 nm) that do not necessarily correlate with the electrochemically active surfaces areas (2.59 to 21.87 $m^2 \, g^{-1}$) of the particles. Some of the $Pt/TiO_2/C$ electrocatalysts showed higher activity toward ORR in sulfuric acid at room temperature than the Pt/C control. Also, a number of the TiO_2-doped electrocatalysts have lower charge transfer resistances, as measured by AC impedance, than the Pt/C control. Cyclic voltammetry shows the hydrogen desorption potential decreases and the potential for reduction of platinum oxide increases in the presence of the added TiO_2. When the performance parameters of the various $Pt/TiO_2/C$ electrocatalysts are compared, it is found the $Pt/TiO_2/C$ electrocatalyst prepared by depositing hydrated TiO_2 onto a Pt/C substrate and subsequently heat treating at 500°C performed best. When integrated into the cathode of a DMFC, the TiO_2 doped electrocatalyst shows higher tolerance to methanol crossover than the Pt/C control. The tolerance becomes more pronounced as the methanol concentration of the fuel stream is raised above 1.0 M. The performance of this TiO_2-doped electrocatalyst, used either in the heat-treated or as prepared form in the cathode of a DMFC was higher than an equivalent cell made with a Pt/C cathode.

There is independent evidence that MWNTs are a carbon support for DMFCs superior to the ubiquitous Vulcan XC-72, and that alloying Pt with other transition metals such as Ni, Fe, and Co produces catalysts with higher activity toward ORR than Pt alone. In a brief and preliminary paper, Li et al. bring these two notions together and test a PtFe alloy supported upon MWNTs for use as an electrocatalyst in DMFC cathodes [50]. A modified polyol strategy was used to prepare the electrocatalyst. Characterization of the material showed that the Pt:Fe ratio was ~1:1; however, the Pt and Fe did not form a stable alloy. The electro-chemical performance of the material was tested subsequent to introduction into the cathodes of DMFCs. It was found the mass activity (current per mg Pt) of the PtFe/MWNT cell held at 600 mV (the activation-controlled cell potential region) was ~40 % higher (4.7 vs. 3.3 mA mg^{-1} Pt) than the Pt/MWNT control cell. At a current density of 300 mA cm^{-2}, the PtFe/MWNT cell held a cell potential of 210 mV versus 151 mV for the Pt/MWNT control. Though the mean particle size for Pt in the PtFe/MWNT material was larger than that of the Pt/MWNT material, the specific activity of the PtFe/MWNT electrocatalyst was more than 2-fold higher at 117 mA m^{-2} Pt versus 50 mA m^{-2} Pt for the Pt/MWNT control. The authors propose the presence of Fe in the material as being respon-sible for the enhanced ORR activity and DMFC performance and that detailed work be conducted to elucidate the Fe and Pt interaction resulting in the enhanced performance.

MEMBRANE TECHNOLOGY

Nafion has been the workhorse PEM of choice for PEFCs and DMFCs for the past 20 years. Its structure is shown in Figure 9.4. While well-suited for use in PEFCs run on hydrogen and oxygen, Nafion is not well-suited for use in DMFCs in large part due to methanol permeability. Efforts to develop a more appropriate PEM for use in DMFCs continues. The ideal membrane is impermeable to methanol, allows facile proton conduction, has good ionic conductivity, can operate over a wide variety of temperatures (e.g., >100°C), and is mechanically and chemically robust. Efforts to develop PEMs appropriate for DMFCs fall roughly into two categories, one focused with developing entirely new PEM materials, the other focused on tailoring the properties of Nafion [24,55,56].

$$\left[\begin{array}{c} [(CF_2CF_2)n\text{-}CF_2CF\text{-}] \\ \qquad (OCF_2CF\text{-})m \; OCF_2CF_2SO_3H \\ \qquad\quad CF_3 \end{array} \right]_x$$

FIGURE 9.4 Chemical structure of Nafion where m is usually 1 and n varies from 6 to 14.

Nafion-Based Membranes

Strategies for tailoring the properties of Nafion membranes include surface modification of the membrane with a barrier less permeable to methanol that still
allows facile proton conduction, addition of intercalants into the membrane to
react with methanol to reduce crossover, and blending Nafion with other polymers
to form hybrid membranes.

Palladium metal is of particular interest for researchers as it is impermeable
to methanol but not proton [33,56]. A variety of methods to apply the Pd layer
to Nafion have been assessed and the effectiveness of the modification in reducing
methanol crossover evaluated.

Ma et al. looked at the effectiveness of reducing methanol crossover by
sputtering a Pt/Pd-Ag/Pt layer onto Nafion [32]. It was found that while the layers
were not crack-free, the Pd alloy-coated Nafion had increased performance over
uncoated Nafion. A DMFC made with a Nafion membrane coated with a 1-μm
thick Pd alloy layer generated a maximum current density roughly 35% (2.3 vs.
1.7 mW cm^{-2}) higher than a control DMFC made with twice the electrocatalyst
loading, though, the power densities are lower than one might expect.

Hejze et al. evaluated the performance of electrolessly deposited Pd-coated
Nafion 117 to that of unmodified Nafion 117 membrane [56]. Here, Pd layers
were coated onto fully hydrated Nafion substrates and the performance as a
separator evaluated in a specialized cell for methanol concentrations <10%. Over
the course of 10 hours, methanol crossover through the Pd-coated membrane was
found to be much lower than for the unmodified Nafion control. The design of
the fuel cell and large iR drop disallowed using realistic current densities. The
authors think the technology can be applied to MEA type cells and acknowledge
the increased cost of using another noble metal in a DMFC.

Methods to deposit Pd onto the surface of Nafion membranes by means of
ion exchange and chemical reduction were studied by Hong et al. [33]. Palladinized Nafion PEMs were found to be less permeable to methanol and uptake
more water then unmodified Nafion. A DMFC made with the palladinized Nafion
PEM generated roughly 40% higher maximum current density than the Nafion
control cell. Of the palladinized membranes tested, the best performers were
found to have nanoparticles deposited near the surface of the membrane.

Choi et al. modified Nafion membranes in one of three ways: plasma etching
the Nafion surface, sputtering Pd onto the Nafion surface, and combining both
plasma etching and Pd sputtering [34]. Plasma etching changes the membrane
morphology. Sputtering Pd "plugs" the pores of the water-rich domain of the
membrane that allow proton conduction, but acts as a barrier to methanol crossover. When tested against an untreated Nafion membrane, it was found that all
three treated membranes had lower permeability to methanol with the plasma/Pd
sputtered membrane having the lowest permeability. When the membranes were
tested in single DMFCs, it was found all of the test cells had higher open circuit
potentials than the control cell. The current-voltage performance of the Pd sputtered and plasma-etched cells were superior to the control cell. The test cell with

both treatments performed poorly relative to the control cell. The Pd sputtered cell had a power maximum of ~80 mW cm^{-2}, which is slightly more than twice the control cell power maximum.

Kim et al. used supercritical carbon dioxide (sCO$_2$) to graft polystyrene onto Nafion 115 as sCO$_2$ produces low thermal stress and has a plasticizing effect when used as a swelling agent in polymers [35]. Following impregnation, the membrane was sulfonated and its properties compared to unmodified Nafion. Impregnated membranes have higher ion exchange capacity and lower permeability to methanol. DMFCs made with impregnated membranes generate more current at 350 mV (~140 mA cm^{-2}) than a Nafion 115 control cell (~113 mA cm^{-2}).

Kang et al. deposited thin (0.1 μm), clay-nanocomposite films onto Nafion 117 using layer-by-layer assembly [57]. The purpose was to reduce methanol crossover using exfoliated (leaf-like) clay nanosheets that are efficient components in barrier membranes for gas and water vapor. Permeability of methanol and ionic conductivity of the treated Nafion are measured and compared to a Nafion control. The control membrane has a methanol permeability of 1.91 × 10^{-6} cm^2 s^{-1} and in-plane conductivity of 0.122 S cm^{-1}. The Nafion modified with a 20-bilayer nanocomposite, has roughly half the methanol permeability as the control Nafion, 7.58 × 10^{-7} cm^2 s^{-1} and nearly the same in-plane ionic conductivity, 0.124 S cm^{-1}.

Chan et al. modified Nafion 115 membranes using *in situ* acid-catalyzed polymerization of furfuryl alcohol (PFA) to introduce highly cross-linked and methanol impermeable domains into the Nafion matrix. Modified and untreated Nafion PEMs were prepared and characterized [58]. It was found that methanol flux through the membranes, measured potentiostatically, changed as a function of the wt% of PFA in the membranes. A "sweet spot" of 8 wt% showed the lowest methanol flux, nearly 3× smaller than unmodified Nafion, while membranes with lower and higher PFA content had methanol flux intermediate to the 8 wt% and control. The PFA membranes were integrated into DMFCs and the output at room temperature and 60°C compared to that of a native Nafion control DMFC. Under all conditions, the DMFCs made with PFA PEMs generated significantly more power than the control DMFC. Peak power densities for DMFCs made with the PFA membranes were 2 to 3× larger than the unmodified Nafion DMFC.

One strategy to enhance water retention in PEFCs is to incorporate inorganic particulates into the PEMs of fuel cells [59,60,61]. Nafion membranes impregnated in this fashion act as a barrier to methanol crossover and can be used in high temperature (~150°C) direct alcohol fuel cells and H$_2$/air fuel cells [44]. Aricò et al. evaluated the surface properties of basic and neutral alumina, ZrO$_2$, SiO$_2$, and SiO$_2$-phosphotungstic acid (SiO$_2$-PWA) using X-ray photoelectron spectroscopy (XPS), Brunauer Emmett and Teller (BET) surface area, and acid-base characterizations. Composite membranes were prepared by recasting Nafion with the inorganic fillers. The resulting membranes were incorporated into DMFCs [44]. All of the membranes showed similar methanol crossover behavior of 4 ± 1 × 10^{-6} mol min^{-1} cm^{-2} at 145°C and 0.5 A cm^{-2}. The electrochemical performance and conductivity of the composite membranes tracks the acidity of the intercalants and follow the series: SiO_2-PWA > SiO_2 > ZrO_2 > n-Al_2O_3 > b-Al_2O_3. That is, the membranes with the best performance were the most acidic. Here "n" stands for neutral and "b" for basic. The DMFC made with the hybrid SiO_2-PWA/Nafion membrane produced 400 mW cm^{-2} at 900 mV using pressurized O_2 and a cell temperature of 145°C.

In another study Baglio et al. studied Nafion-TiO_2 membranes for use in high temperature DMFCs [45]. The electrochemical performance of the membrane is significantly influenced by the properties of the intercalants such as surface area and pH. Here, the TiO_2 intercalated into the membranes had been calcined at temperatures ranging between 500 and 800°C. The pH of the resulting particles trends with the temperature; a lower calcination temperature gives a particle with lower pH. As seen in [44], the membrane with the highest acidity generated the most power. Under similar operating conditions of 145°C cell temperature, 2.0 M methanol solution, and pressurized O_2, a DMFC using TiO_2 calcined at 500°C generated 350 mW cm^{-2}. This DMFC, subjected to longevity tests, operated for a month with daily cycles of start-up and shut-down under the conditions mentioned above. The cell potential was potentiostatically held at 400 mV. At start-up, the cell generated ~800 mA cm^{-2} (320 mW cm^{-2}) and decreased to roughly 760 mA cm^{-2} (304 mW cm^{-2}) after 4 hours of operation. This occurred daily. Upon feeding water to the anode, start-up performance was restored the following day.

Bauer and Willert-Porada characterized Zr-phosphate-Nafion membranes as candidate materials for use in DMFCs [62]. The inorganic filler reduced methanol permeability and the phosphate layer had preferred permeability to water over methanol. The preliminary results suggest Zr-phosphate can be used to tailor such Nafion properties.

New Separators

New materials are being developed for use as separators/ionic conductors in DMFCs. These materials generally fall into one of two categories, fluorinated and nonfluorinated. Here some current developments in both areas are presented.

In an effort to develop an inexpensive and effective separator for DMFCs and fuel cells operated on other fuels, Melman et al. developed a nanoporous proton-conducting membrane (NP-PCM) [63]. The NP-PCM is made of polyvinylidene fluoride (PVDF) and SiO_2. Fuels are mixed into 3 M sulfuric acid electrolyte and circulated past the anode. Characteristics of the NP-PCM separator that bests Nafion include: membrane cost lowered by 2 orders of magnitude; pore sizes roughly 50% that of Nafion; methanol crossover cut in half; as much as 4× greater ionic conductivity; and a membrane insensitive to heavy metal corrosion products that allow for less expensive hardware and catalysts. The maximum power achieved with the cell was 85 mW cm^{-2} at ~243 mV on oxygen at atmospheric pressure, and cell temperature of 80°C. One disadvantage of this type of membrane is the peripheral systems must be corrosion resistant.

Nafion is made by copolymerizing tetrafluoroethylene (TFE) and perfluorovinyl ether (PSEPVE) containing a sulfonyl fluoride. Whereas Nafion provides good performance in PEFCs at temperatures below 100°C, the material is expensive due to the relatively difficult polymerization and expense of the monomer PSEPVE. Also, chemical modifications to tailor Nafion properties has proven difficult. Yang and Rajendran describe an effective copolymerization strategy of TFE and ethylene to produce less expensive melt-processable terpolymers that are easily hydrolyzed and acidified to give polymers of high ionic conductivity [42]. The conductivity is similar to or slightly higher than that of Nafion. The membrane also uptakes more water (by weight) than Nafion under the same conditions. An unoptimized 5-mil thick membrane tested in a DMFC yielded performance comparable to a DMFC made with Nafion 115. The power output of the DMFCs were quite similar, though the methanol crossover through the experimental membrane is 9% higher than for Nafion 115 (i.e., 10.9×10^{-4} vs. 9.9×10^{-4} g min^{-1} cm^{-2}). The authors speculate that the PEM may be optimized by making a more homogeneous thickness and cross-linking.

A review of nonfluorinated PEMs for use in DMFCs was prepared by Rozière and Jones [38]. The number of nonfluorinated polymer materials for application in higher-temperature fuel cells (i.e., > 80°C) is limited by thermal instability. Thermally stable polymers tend to have either polyaromatic or polyheterocyclic repeat units. Examples of these include polybenzimidazole (PBI), poly(ether ketone)s (PEK), poly(phenyl quinoxaline (PPQ), polysulfone (PSU), and poly(ether sulfone) (PES). The chemical structures of some common nonfluorinated polymers are shown in Figure 9.5. These polymers are thermally stable but are poor ionic conductors until modified. Modification methods include acid and base doping of the polymer, sulfonation of the polymer backbone, grafting phosphonated or sulfonated functional groups onto the polymer (where sulfonated polymers generally contain "s" in the acronym), graft polymerization onto the polymer followed by functionalization of the graft material, and total synthesis. Many of these membranes exhibit good proton conductivity (0.01 to 0.1 S cm^{-1}) when well-hydrated but a balance between the sulfonation level imparting ionic conductivity, adequate membrane strength, and membrane swelling must be met before long-term use in fuel cells is possible. Rozière and Jones suggest that sPEKs and related polymer blends hold the most promise for use in DMFCs. They cite unpublished performance data of a DMFC made using polyaromatic membrane with comparable power density to Nafion cells and little degradation after start/stop regimes at temperatures >100°C over a period of weeks.

There is a limited amount of DMFC performance data for fuel cells made with these PEMs, and the conditions under which the PEMs are tested in DMFCs are quite varied. Relevant DMFC performance data included here are listed in the "Performance Targets and Efficiencies" section of this chapter. A more rigorous set of evaluations for candidate PEMs in operational DMFCs should include durability tests including startup and shutdowns over extended periods.

An engaging two-paper study by Silva et al. evaluates inorganic-organic hybrid membranes made from sulfonated poly(ether ether ketone) (sPEEK)

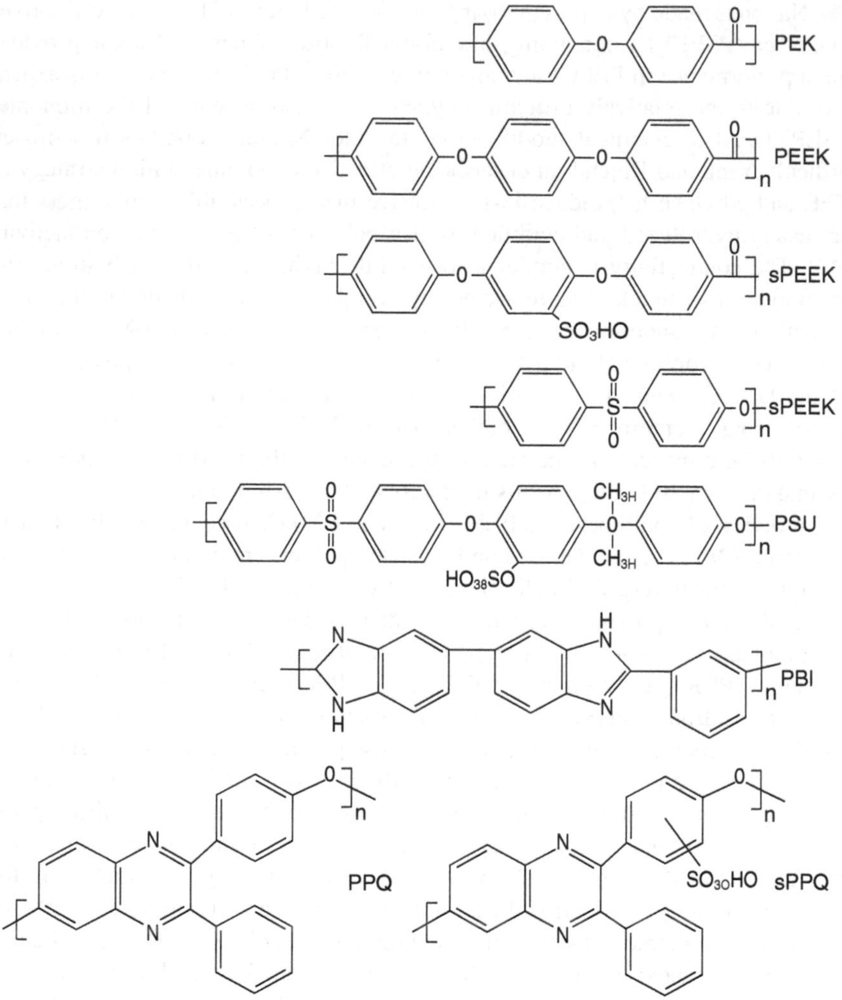

FIGURE 9.5 Chemical structures of some of the more common nonfluorinated polymers being tested as PEMs. Structures redrawn from [38].

[64,65]. One goal of the study is to demonstrate a systematic and complete approach to DMFC membrane development, characterization, and, ultimately, the real-world testing in DMFCs that is thus far lacking for other promising materials. The sPEEK membranes have a sulfonation degree of 87% and zirconium oxide content that varies between 2.5 and 12.5 wt%. The group evaluates the membranes using standard analytical techniques (e.g., impedance spectroscopy (proton conductivity and proton transport resistance), pervaporation (permeability to methanol), and water swelling) and as PEMs in DMFCs. The organic/inorganic hybrid sPEEK/ZrO$_2$ membranes exhibit good proton conductivity and the addition of ZrO$_2$ particles can tailor the electrochemical performance of the membranes. This

makes the organic/inorganic hybrid sPEEK/ZrO_2 membranes a possible alternative to perfluorinated membranes. It was found the proton transport resistance increased as the wt% ZrO_2 in the membrane increased and that proton conductivity followed the opposite trend. Water uptake decreases as the inorganic component increases, following the same trend as the proton conductivity. This supports the observation of the importance of sorbed water in proton conduction in sulfonated matrices. Methanol permeability decreases as the inorganic content of the membrane increases. The permeability of O_2, CO_2, and N_2, reduction products and oxidant stream constituents were evaluated as a function inorganic content of the membrane. It was found that as ZrO_2 content increases, the permeability of O_2 and CO_2 decreases and N_2 permeability is unaffected. In general, the desirable characteristic of decreasing permeability of reactants, products, and oxidant stream constituents as the ZrO_2 content increases were observed but with a corresponding decrease in proton conductivity.

Jörissen et al. evaluated DMFCs made with nonfluorinated PEMs (sPEEK and sPEK using PBI and basically substituted PSU (bPSU)) prepared by a thin-film method [24], in which the catalytic layer is sprayed directly onto the PEM. The authors used unsupported electrocatalysts at relatively high loadings (1.5 to 6.5 mg cm^{-2}) to overcome the effects of flooding at both electrodes. The fuel cells were operated at 110°C with little or no cathode humidification and performance was compared to DMFCs made with Nafion 117 and 105. None of the DMFCs prepared from the test PEMs had optimized catalyst layer/PEM interfaces, yet a number of the test cells performed similarly to the control cell made with Nafion 105. The best performing of these cells was a membrane composed of PEK, PBI, and bPSU that generated 85% of the power density (230 mW cm^{-2} @ 500 mV) of the Nafion 105 cell (270 mW cm^{-2} @ 500 mV). The control cell made from Nafion 117 exhibited poor performance generating less than half the power density of Nafion 105 cell under the same conditions.

Another nonfluorinated material of promise is poly(vinyl alcohol) (PVA). The material is chemically and thermally robust and quite inexpensive relative to Nafion. In a recent article by Khan et al., the synthesis and characterization of PVA-based membrane is described [66]. Membranes were based on PVA and its ionic blends with sodium alginate (SA) and chitosan (CS). The membrane ion exchange capacity (IEC, in milliequivalents of ion per gram dry polymer) were determined to be ~0.5 for PVA, 0.6 for PVA-CS and 0.8 for PVA-SA. All of the PVA-based membranes were found to have lower methanol permeability than Nafion. The PVA-CS membrane had the lowest permeability at 6.9×10^{-8} cm^2 s^{-1} as compared to Nafion 117 with a permeability of 2.76×10^{-7} cm^2 s^{-1}. The other two membranes had permeability intermediate to PVA-SA and Nafion. Contrasting the desirable characteristic of lower methanol permeability than Nafion 117, the proton conductivities of the PVA-based membranes are significantly lower (~0.01 S cm^{-1}) than that for Nafion 117 (0.1 S cm^{-1}). The investigators hold out the possibility of improving proton conduction in the PVA membranes through doping.

Lee et al. also investigated the use of PVA-based materials. The group prepared membranes made from PVA, SiO_2, and sulfosuccinic acid (SSA) [67]. The SSA acted both as the cross-linking agent and as the source of hydrophilic sulfonate groups. The SSA content of the membranes was varied from 5 to 25 wt %. Both the proton conductivity and methanol permeability decreased as the wt % of SSA increased to 20 wt %. At over 20 wt % SSA both trends reversed. Similar to what Khan et al. found, the membranes had proton conductivities on the order of 1×10^{-3} to 1×10^{-2} S cm^{-1} and methanol permeability of 1×10^{-8} to 1×10^{-7} cm^2 s^{-1}. In a related study, Lee and his group expanded the previous work by adding poly(acrylic acid) (PAA) to the hybrid membranes [68]. The thought was the addition of unreacted carboxylic acid groups may improve proton conductivity, but the PAA did not markedly improve upon the original hybrid membranes.

CLOSING THOUGHTS

A general observation made while surveying the DMFC literature is the unsettling lack of replicates in many studies. Frequently, one test cell is compared to one control cell and no statistics are provided to gauge cell performance. The incomplete reporting of test condtions further confounds comparison. While impractical for some studies, an analysis of cell performance over extended periods of time and with intermittent startup and shutdowns should be the standard. As a result of the limited data, broad conclusions drawn from these studies must be assessed with caution.

As can be seen from the breadth of topics covered in this chapter, many avenues are being pursued to make DMFC technology a practical reality. Effort is expended in the engineering of hardware and fuel delivery systems and in developing PEMs impermeable to methanol. That being said, the outlook for DMFC technology appears mixed. At this point, the development of alternative (e.g., less expensive, more effective, more robust) electrocatalysts appears to be the foremost obstacle in making DMFCs a practicable power generation alternative.

An important study by Zelenay et al. shows that use of the highly active PtRu electrocatalyst present in the great majority of DMFC anodes (see Tables 9.2 and 9.3) will contaminate the cathode with Ru [43]. The migration of Ru occurs under nearly all operating conditions, including conditions where the cell is only humidified with inert gases and no current is drawn from the cell. Cathode contamination with Ru inhibits oxygen reduction kinetics and reduces cathode electrocatalyst tolerance to methanol crossover. The degree to which Ru contamination occurs depends on such factors as anode potential and operating lifetime of the cell. Depending upon the degree of Ru contamination at the cathode, the associated loss of cell performance is estimated to be from as little as ~40 mV to as much as 200 mV. Figure 9.6 shows data for a Pt-only cathode subject to Ru contamination.

The top plots of Figure 9.6, marked "a" and "b," include CO stripping scans (a) and cyclic voltammograms recorded in the absence of CO (b) for a DMFC cathode made with Pt only as the electrocatalyst. The DMFC cathode was part

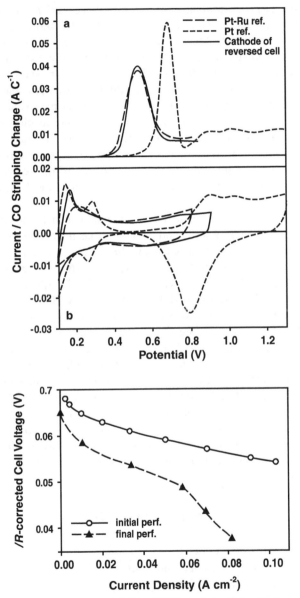

FIGURE 9.6 The top plots (marked "a" and "b") include CO stripping scans (a), and cyclic voltammograms recorded in the absence of CO (b), for a DMFC cathode made with Pt only as the electrocatalyst. The solid black line denotes the DMFC cathode. The control electrodes, Pt only (short dashed lines) and PtRu (long dashed lines) are also shown. The bottom plot is the iR corrected voltage-current plot for the DMFC cathode. The cell was fed a 0.3-M methanol solution, cell temperature is 70°C, dry air is the oxidant, and no backpressure is applied at the cathode. The solid line plots the initial performance of the cell and the dashed line plots the performance of the cell after inclusion in the DMFC stack for 6 months. Reproduced by permission of the Electrochemical Society, Inc.

of a 22-cell stack run intermittently for 6 months that experienced voltage reversal. The plots include data for control electrodes, Pt only (short dashed lines) and PtRu (long dashed lines). As can be seen in (a), the Pt-only DMFC cathode behaves in a near identical fashion to the PtRu control, consistent with PtRu in the electrocatalyst layer of the cathode. The CVs of plot (b) shows a similar trend with the traces for the DMFC cathode morphologically similar to the PtRu control.

The bottom plot of Figure 9.6 shows how cell performance can diminish with Ru migration. The plot is the *iR* corrected voltage-current plot for the DMFC cathode described above. The cell was fed a 0.3-M methanol solution, cell temperature is 70°C, dry air is the oxidant, and no backpressure is applied at the cathode. The solid line plots the initial performance of the cell and the dashed line plots the performance of the cell after inclusion in the DMFC stack for 6 months.

It should be noted the fate of the cathode shown in Figure 9.6 is a rather extreme example of Ru migration and the conditions the cell experienced are not ideal, however, as the researchers found, Ru migration occurs even under the most benign conditions. Similar electrocatalyst migration in the form of agglomeration was found by Yi et al. following a 75-hour DMFC life test [29]. The PtRu electrocatalyst used by Yi differed from the Zelenay study in that the anode catalyst is carbon supported PtRu rather than PtRu black and suggests that all forms of PtRu electrocatalysts are likely susceptible to migration in DMFC.

In summary, the principle challenges to commercializing DMFC technology are effective and stable electrocatalysts tolerant to methanol contamination of the cathode and membranes significantly less permeable to methanol. The most commonly used separator, Nafion, is not sufficient for the task. Nor are Pt and PtRu electrocatalysts. The outlook for membrane development for the near future seems to be somewhat static. However, promising electrocatalyst developments such as relatively inexpensive Pd-based catalysts that are stable and reaction specific may be the breakthrough that allows for the use of a less than ideal separator such as Nafion.

ABBREVIATIONS

AFC	Alkaline Fuel Cell
BET	Brunauer Emmett and Teller
BOL	Beginning of Life
CS	Chitosan
DARPA	Defense Advanced Research Projects Agency
PPQ	Poly(Phenyl Quinoxaline)
DMFC	Direct Methanol Fuel Cell
EDX	Energy Dispersive X-ray
ELAT	Commercial Gas Diusion Electrode (E-Tek)
PVA	Poly(Vinyl Alcohol)
EOL	End of Life

IEC	Ion Exhcange Capacity
J*max*	Maximum Current Density
MCFC	Molten Carbonate Fuel Cell
MOR	Methanol Oxidation Reaction
MWNT	Multiwall Carbon Nanotubes
NP-PCM	Nanoporous Proton-Conducting Membrane
SSA	Sulfosuccinic Acid
ORR	Oxygen Reduction Reaction
OTC	Operational Test Command
PAFC	Phosphoric Acid Fuel Cell
PBI	Polybenzimidazole
PEFC	Polymer Electrolyte Fuel Cell
PEK	Poly(Ether Ketone)
PEM	Polymer Electrolyte Membrane
PES	Poly(Ether Sulfone)
PFA	Polymerized Furfuryl Alcohol
PSEPVE	Perfluorovinyl Ether
PSU	Polysulfone
SA	Sodium Alginate
SCE	Saturated Calomel Electrode
SEM	Scanning Electron Microscopy
SiO_2-PWA	SiO_2-Phosphotungstic Acid
SOFC	Solid Oxide Fuel Cell
sPEEK	Sulfonated Poly(Ether Ether Ketone)
SWNT	Single-Wall Carbon Nanotubes
TEM	Transmission Electron Microscopy
TFE	Tetrafluoroethylene
XPS	X-ray Photoelectron Spectroscopy
XRD	X-ray Diffraction Analysis

REFERENCES

1. McNicol, B.D., Rand, D.A.J. and Williams, K.R., Direct Methanol-Air Fuel Cells for Road Transportation, *Journal of Power Sources*, 83, 15–31, 1999.
2. Perry, M.L. and Fuller, T.F., A Historical Perspective of Fuel Cell Technology in the 20th Century, *Journal of the Electrochemical Society*, 149, S59–S67, 2002.
3. Stone, C. and Morrison. A.E., From Curiosity to "Power to Change the World," *Solid State Ionics*, 152, 153, 1–13, 2002.
4. Glazebrook, R.W., Sciences of Heat Engines and Fuel Cells: The Methanol Fuel Cell as a Competitor to Otto and Diesel Engines, *Journal of Power Sources*, 7, 215–256, 1982.
5. Surampudi, S., Narayanan, S.R., Vamos, E., Frank, H., Halpert, G., LaConti, A., Kosek, Prakash, G.K.S. and Olah, G.A., Advances in Direct Oxidation Methanol Fuel Cells, *Journal of Power Sources*, 47, 377–385, 1994.

6. Costamagna, P. and Srinivasan, S., Quantum Jumps in the PEMFC Science and Technology from the 1960s to the Year 2000 Part I. Fundamental Scientific Aspects, *Journal of Power Sources*, 102, 242–252, 2001.

7. Bagotskii, V.S. and Vasil'ev, Y.B., Mechanism of Electrooxidation of Methanol on the Platinum Electrode, *Electrochimica Acta* 12, 1323–1343, 1967.

8. Petry, O.A., Podlovchenko, B.I., Frumkin, A.N. and Lal, H., The Behavior of Platinized-Platinum and Platinum-Ruthenium Electrodes in Methanol Solutions, *Journal of Electroanalytical Chemistry*, 10, 253–269, 1965.

9. Verbrugge, M.W., Methanol Diusion in Perfluorinated Ion-exchange Membranes, *Journal of the Electrochemical Society*, 136, 417–423, 1989.

10. Aramata, A. and Ohnishi, R., Methanol Electrooxidation on Platinum Directly Bonded to a Solid Polymer Electrolyte Membrane, *Journal of Electroanalytical Chemistry and Interfacial Electrochemistry*, 162, 153–162, 1984.

11. Katayama-Aramata, A. and Ohnishi, R., Metal Electrodes Bonded on Solid Polymer Electrolytes: Platinum Bonded on Solid Polymer Electrolyte for Electrooxidation of Methanol in Perchloric Acid Solution, *Journal of the American Chemical Society*, 105, 658, 659, 1983.

12. Watanabe, M., Takahashi, T. and Kita, H., The Use of Thin Films of Sulfonated Fluoropolymers for Improvements in the Activity and Durability of Platinum Electrocatalysts for Methanol Electrooxidation, *Journal of Electroanalytical Chemistry and Interfacial Electrochemistry*, 284, 511–515, 1990.

13. Apanel, G., Direct Methanol Fuel Cells—Ready to go Commercial?, *Fuel Cells Bulletin*, Nov., 12–17, 2004.

14. Dunwoody, D., Gellett, W., Chung, H. and Leddy, J., Magnetic Modification of Proton Exchange Membrane Fuel Cells for Improved Carbon Monoxide Tolerance, in *40th Power Sources Conference*, Cherry Hill, NJ, 2002, pp. 262–265.

15. Gellett, W.L., Magnetic Microparticles on Electrodes: Polymer Electrolyte Membrane Fuel Cells, Carbon Monoxide Oxidation, and Transition Metal Complex Electrochemistry, Ph.D. thesis, University of Iowa, 2004.

16. Leddy, J., Gellett, W.L. and Dunwoody, D.C., Self-Hydrating Membrane Electrode Assemblies for Fuel Cells, in *US Patent Application*, USA, Filed Oct. 15, 2003, published 21 April 2005, University of Iowa.

17. Lamy, C. and Léger, J.M., Advanced Electrode Materials for the Direct Methanol Fuel Cell, in *Interfacial Electrochemistry Theory, Experiment, and Applications*, Wreckowski, Ed., Marcel Dekker, Inc., New York, 1999, pp. 885–894.

18. Galus, Z., Carbon, Silicon, Germanium, Tin, and Lead, in *Standard Potentials in Aqueous Solution*, Bard, A.J., Parsons, R. and Jordan, J., Eds., Marcel Dekker, Inc.

19. Harris, D.C., *Quantitative Chemical Analysis*, W.H. Freeman and Company, New York, sixth ed., 2003.

20. Vielstich, W., CO, Formic Acid, and Methanol Oxidation in Acid Electrolytes— Mechanisms and Electrocatalysis, in *Encyclopedia of Electrochemistry, Volume 2: Interfacial Kinetics and Mass Transport*, Bard, A.J., Stratmann, M. and Calvo, E.J., Eds.

21. Hamnett, A., Mechanism of Methanol Electro-Oxidation, in *Interfacial Electrochemistry: Theory, Experiment, and Applications*, Wieckowski, A., Ed., Marcel Dekker, Inc.

22. Parsons, R. and VanderNoot, T., The Oxidation of Small Organic Molecules: A Survey of Recent Fuel Cell Related Research, *J. Electroanal. Chem.*

23. Gasteiger H.A., Markovic, N., P. N. Ross Jr., and Cairns, E.J., Electro-Oxidation of Small Organic Molecules on Well-Characterized Pt-Ru Alloys, *Electrochimica Acta*, 39, 1825–1832, 1994.

24. Jorissen, L., Gogel, V., Kerres, J. and Garche, J., New Membranes for Direct Methanol Fuel Cells, *Journal of Power Sources*, 105, 267–273, 2002.

25. Electrochemical Technologies Group, Jet Propulsion Laboratory, *Additional Information Direct Methanol Fuel Cell Requirement*, 2005, https://www.datm01.atec.army.mil/readroom%20DMFC.htm

26. Electrochemical Technologies Group, Jet Propulsion Laboratory, *300-Watt Power Source Development at the Jet Propulsion Laboratory*, 2005, https://www.datm01.atec.army.mil/readroom%20DMFC.htm.

27. Knights, S.D., Colbow, K.M., St-Pierre, J. and Wilkinson, D.P., Aging Mechanisms and Lifetime of PEFC and DMFC, *Journal of Power Sources*, 127, 127–134, 2004.

28. Ball Aerospace & Technologies Corp., Direct-methanol Fuel Cell—20 Watts Preliminary Technology Update, 2005, http://www.ballaerospace.com/pdf/pps20.pdf.

29. Liu, J., Zhou, Z., Zhao, X., Xin, Q., Sun, G. and Yi, B., Studies on Performance Degradation of a Direct Methanol Fuel Cell (DMFC) in Life Test, *Physical Chemistry Chemical Physics* 6, 134–137, 2004.

30. Ge, J. and Liu, H., Experimental Studies of a Direct Methanol Fuel Cell, *Journal of Power Sources*, 142, 56–69, 2005.

31. Jung, D.H., Lee, C.H., Kim, C.S. and Shin, D.R., Performance of a Direct Methanol Polymer Electrolyte Fuel Cell, *Journal of Power Sources* 71, 169–173, 1998.

32. Ma, Z.Q., Cheng, P. and Zhao, T.S., A Palladium-Alloy Deposited Nafion Membrane for Direct Methanol Fuel Cells, *Journal of Membrane Science*, 215, 327–336, 2003.

33. Kim, Y.J., Choi, W.C., Woo, S.I. and Hong, W.H., Evaluation of a palladinized Nafion for direct methanol fuel cell application, *Electrochimica Acta*, 49, 3227–3234, 2004.

34. Choi, W.C., Kim, J.D. and Woo, S.I., Modification of Proton Conducting Membrane for Reducing Methanol Crossover in a Direct-Methanol Fuel Cell, *Journal of Power Sources*, 96, 411–414, 2001.

35. Sauk, J., Byun, J. and Kim, H., Grafting of styrene on to Nafion membranes using supercritical CO2 impregnation for direct methanol fuel cells, *Journal of Power Sources* 132, 59–63, 2004.

36. Carmo, M., Paganin, V.A., Rosolen, J.M. and Gonzalez, E.R., Alternative Supports for the Preparation of Catalysts for Low-Temperature Fuel Cells: The Use of Carbon Nanotubes, *Journal of Power Sources*, 142, 169–176, 2005.

37. Furukawa, K., Okajima, K. and Sudoh, M., Structural Control and Impedance Analysis of Cathode for Direct Methanol Fuel Cell, *Journal of Power Sources*, 139, 9–14, 2005.

38. Roziere, J. and Jones, D.J., Non-Fluorinated Polymer Materials for Proton Exchange Membrane Fuel Cells, *Annual Review of Materials Research*. 33, 503–555, 2003.

39. Zhang, T. and Wang, Q.-M., Valveless Piezoelectric Micropump for Fuel Delivery in Direct Methanol Fuel Cell (DMFC) Devices, *Journal of Power Sources*, 140, 72–80, 2005.

40. Gurau, B. and Smotkin, E.S., Methanol Crossover in Direct Methanol Fuel Cells: A Link Between Power and Energy Density, *Journal of Power Sources*, 112, 339–352, 2002.

41. Allen, R.G., Lim, C., Yang, L.X., Scott, K. and Roy, S., Novel Anode Structure for the Direct Methanol Fuel Cell, *Journal of Power Sources*, 143, 142–149, 2005.

42. Yang, Z.-Y. and Rajendran, R.G., Copolymerization of Ethylene, Tetrafluoroethylene, and an Olefin-containing Fluorosulfonyl Fluoride: Synthesis of High-Proton-Conductive Membranes for Fuel-Cell Applications, *Angewandte Chemie, International Edition*, 44, 564–567, S564/1–S564/5, 2005.

43. Piela, P., Eickes, C., Brosha, E., Garzon, F. and Zelenay, P., Ruthenium Crossover in Direct Methanol Fuel Cell with Pt-Ru Black Anode, *Journal of The Electrochechemical Society*, 151, A2053–A2059, 2004.

44. Arico, A.S., Baglio, V., Blasi, A.D., Modica, E., Antonucci, P.L. and Antonucci, V., Surface Properties of Inorganic Fillers for Application in Composite Membranes—Direct Methanol Fuel Cells, *Journal of Power Sources*, 128, 113–118, 2004.

45. Baglio, V., Arico, A.S., Blasi, A.D., Antonucci, V., Antonucci, P.L., Licoccia, S., Traversa, E. and Fiory, F.S., Nafion-TiO2 Composite DMFC Membranes: Physicochemical Properties of the Filler Versus Electrochemical Performance, *Electrochimica Acta*, 50, 1241–1246, 2005.

46. Yu, H.B., Kim, J.-H., Lee, H.-I., Scibioh, M.A., Lee, J., Han, J., Yoon, S.P. and Ha, H.Y., Development of Nanophase CeO_2-Pt/C Cathode Catalyst for Direct Methanol Fuel Cell, *Journal of Power Sources*, 140, 59–65, 2005.

47. Zhou, W.J., Zhou, B., Li, W.Z., Zhou, Z.H., Song, S.Q., Sun, G.Q., Xin, Q., Douvartzides, S., Goula, M. and Tsiakaras, P., Performance Comparison of Low-Temperature Direct Alcohol Fuel Cells with Different Anode Catalysts, *Journal of Power Sources*, 126, 16–22, 2004.

48. Xiong, L. and Manthiram, A., Synthesis and Characterization of Methanol Tolerant Pt/TiOx/C Nanocomposites for Oxygen Reduction in Direct Methanol Fuel Cells, *Electrochimica Acta* 49, 4163–4170, 2004.

49. Li, H., Xin, Q., Li, W., Zhou, Z., Jiang, L., Yang, S. and Sun, G., An Improved Palladium-Based DMFCs Cathode Catalyst, *Chemical Communications*, Cambridge, U.K., 2776–2777, 2004.

50. Li, W., Liang, C., Qiu, J., Li, H., Zhou, W., Sun, G. and Xin, Q., Multi-Walled Carbon Nanotubes Supported Pt-Fe Cathodic Catalyst for Direct Methanol Fuel Cell, *Reaction Kinetics and Catalysis Letters*, 82, 235–240, 2004.

51. Bockris, J.O. and Srinivasan, S., *Fuel Cells: Their Electrochemistry*, McGraw-Hill, New York, 1969.

52. Kulikovsky, A.A., Model of the Flow with Bubbles in the Anode Channel and Performance of a Direct Methanol Fuel Cell, *Electrochemistry Communications*, 7, 237–243, 2005.

53. Sarma, L.S., Lin, T.D., Tsai, Y.-W., Chen, J.M. and Hwang, B.J., Carbon-Supported Pt-Ru Catalysts Prepared by the Nafion Stabilized Alcohol-Reduction Method for Application in Direct Methanol Fuel Cells, *Journal of Power Sources*, 139, 44–54, 2005.

54. Shim, J., Lee, C.-R., Lee, H.-K., Lee, J.-S. and Cairns, E.J., Electrochemcial Characteristics of Pt-WO3/C and Pt-TiO2/C Electrocatalysts in a Polymer Electrolyte Fuel Cell, *Journal of Power Sources*, 102, 172–177, 2001.

55. Manea, C. and Mulder, M., New Polymeric Electrolyte Membranes Based on Proton Donor-Proton Acceptor Properties for Direct Methanol Fuel Cells, *Desalination*, 147, 179–182, 2002.

56. Hejze, T., Gollas, B.R., Sauerbrey, R.K., Schmied, M., Hofer, F. and Besenhard, J.O., Preparation of Pd-Coated Polymer Electrolyte Membranes and their Application in Direct Methanol Fuel C, *Journal of Power Sources*, 140, 21–27, 2005.

57. Kim, D.W., Choi, H.-S., Lee, C., Blumstein, A. and Kang, Y., Investigation on Methanol Permeability of Nafion Modified by Self-Assembled Clay-Nanocomposite Multilayers, *Electrochimica Acta* 50, 659–662, 2004.

58. Liu, J., Wang, H., Cheng, S. and Chan, K.-Y., Nafion-polyfurfuryl Alcohol Nanocomposite Membranes for Direct Methanol Fuel Cells, *Journal of Membrane Science*, 246, 95–101, 2005.

59. Watanabe, M., Uchida, H., Seki, Y. and Emori, M., Self-Humidifying Polymer Electrolyte Membranes for Fuel Cells, *Journal of The Electrochemical Society*, 143, 3847–3852, 1996.

60. Watanabe, M., Uchida, H. and Emori, M., Polymer Electrolyte Membranes Incorporated with Nanometer-Size Particles of Pt and/or Metal-Oxides: Experimental Analysis of the Self-Humidification and Suppression of Gas-Crossover in Fuel Cells, *Journal of Physical Chemistry B*, 102, 3129–3137, 1998.

61. Watanabe, M., Uchida, H., Ueno, Y. and Hagihara, H., Self-Humidifying Electrolyte Membranes for Fuel Cells, *Journal of the Electrochemical Society*, 150, A57–A62, 2003.

62. Bauer, F. and Willert-Porada, M., Microstructural characterization of Zr-phosphate-Nafion membranes for direct methanol fuel cell (DMFC) applications, *Journal of Membrane Science*, 233, 141–149, 2004.

63. Peled, E., Duvdevani, T., Aharon, A. and Melman, A., Direct-Oxidation Fuel Cells Based on Nanoporous Proton- Conducting Membrane (NP-PCM), In *Direct Methanol Fuel Cells: Proceedings of the International Symposium*, S. R. Narayanan, S.R., Gottesfeld, S. and Zawodzinski, T., Eds.

64. Silva, V.S., Rumann, B., Silva, H., Gallego, Y.A., Mendes, A., Madeira, L.M. and Nunes, S.P., Proton Electrolyte Membrane Properties and Direct Methanol Fuel Cell Performance I. Characterization of Hybrid Sulfonated Poly (ether ether ketone)/Zirconium Oxide Membranes, *Journal of Power Sources*, 140, 34–40, 2005.

65. Silva, V.S., Schirmer, J., Reissner, R., Rumann, B., Silva, H., Mendes, A., Madeira, L.M. and Nunes, S.P., Proton Electrolyte Membrane Properties and Direct Methanol Fuel Cell Performance II. Fuel Cell Performance and Membrane Properties Eects, *Journal of Power Sources*, 140, 41–49, 2005.

66. Smitha, B., Sridhar, S. and Khan, A.A., Synthesis and characterization of poly(vinyl alcohol)-based membranes for direct methanol fuel cell, *Journal of Applied Polymer Science*, 95, 1154–1163, 2005.

67. Kim, D.S., Park, H.B., Rhim, J.W. and Lee, Y.M., Preparation and Characterization of Crosslinked PVA/SiO2 Hybrid Membranes Containing Sulfonic Acid Groups for Direct Methanol Fuel Cell Applications, *Journal of Membrane Science*, 240, 37–48, 2004.

68. Kim, D.S., Park, H.B., Rhim, J.W. and Lee, Y.M., Proton Conductivity and Methanol Transport Behavior of Cross-Linked PVA/PAA/Silica Hybrid Membranes, *Solid State Ionics*, 176, 117–126, 2005.

10 Direct Ethanol Fuel Cells

Shelley D. Minteer
Department of Chemistry, Saint Louis University, Missouri

CONTENTS

Abstract This chapter details the background and performance of direct ethanol fuel cells (DEFCs). This chapter compares direct ethanol fuel cells to direct methanol fuel cells and other alcohol-based fuel cells. It discusses recent developments in bimetallic electrocatalysts, membrane electrode assembly (MEA) fabrication techniques, temperature effects, and the effects of fuel concentration on the performance of the direct ethanol fuel cell.

INTRODUCTION

Portable power requires simplistic systems that operate at or near room temperature. Most research in fuel cells for use as portable power have employed polymer electrolyte membrane fuel cells. Polymer electrolyte membrane (PEM) fuel cells can be characterized into two categories: reformed and direct systems. Reformed systems require the use of an external reformer to reform a fuel (methane, methanol, ethanol, gasoline, etc.) into hydrogen for use in the fuel cell. In direct systems the fuel is oxidized at the surface of the electrode without treatment. Over the last 40 years, there has been extensive research on direct methanol fuel cells (DMFC) for portable power applications at low to moderate temperatures [1–4]. However, there are a number of problems associated with the use of methanol as a fuel for portable power supplies. Methanol is highly toxic and could lead to long-term environmental problems because methanol is so miscible in water [5]. These limitations have led researchers to investigate other fuels. Ethanol is an attractive alternative to methanol as a fuel for a fuel cell. Ethanol is a renewable fuel and can be produced from farm products and biomass. Ethanol and its intermediate oxidation products have been shown to be less toxic than other alcohols [6]. The problem with ethanol as a fuel (in comparison to methanol) is that complete oxidation of ethanol requires the breaking of a C–C bond, which is difficult at traditional Pt-based catalysts. This typically leads to incomplete oxidation of ethanol, which decreases the efficiency of the fuel cells and could provide toxic by-products or electrode passivation. This chapter focuses on the basics of direct ethanol fuel cells and the effects of catalyst, temperature, and fuel concentration on fuel cell performance.

DIRECT ETHANOL FUEL CELLS

Direct ethanol fuel cells in the literature are PEM-style fuel cells. They contain three main components: the anode, the cathode, and the polymer electrolyte membrane that separates the anode solution from the cathode solution. This polymer electrolyte membrane is a proton transport membrane that is typically a cation-exchange polymer like either Nafion 115 or Nafion 117. Figure 10.1 shows a schematic of the typical direct ethanol fuel cell. Assuming complete oxidation of ethanol, the anode reaction is as follows:

Anode: $CH_3CH_2OH + 3H_2O \rightarrow 2CO_2 + 12H^+ + 12e^-$

where ethanol forms only carbon dioxide as a by-product. The cathode reaction is reduction of oxygen, which is shown below along with the overall reaction for

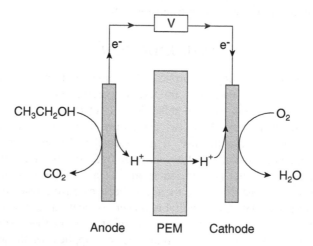

FIGURE 10.1 Schematic of a direct ethanol fuel cell.

the fuel cell. The theoretical open circuit potential for a DEFC is 1.145V [7], which is comparable to direct methanol fuel cells.

Cathode: $O_2 + 4H^+ + 4e^- \rightarrow 2H_2O$

Overall reaction: $CH_3CH_2OH + 3O_2 \rightarrow 2CO_2 + 3H_2O$

A theoretical investigation into the comparison of direct methanol and direct ethanol fuel cells shows that direct ethanol cells have higher theoretical energy densities compared to direct methanol fuel cells. The energy density of a direct ethanol fuel cell is 8.01 kW/kg compared to 6.09 kW/kg for a direct methanol fuel cell [8].

ETHANOL ELECTROCATALYSTS

The major problem associated with using ethanol as a fuel is the low reaction kinetics of ethanol oxidation versus methanol oxidation [9]. Traditional hydrogen/oxygen PEM fuel cells and DMFCs typically employ Pt-based catalysts for oxidation of fuel, but pure Pt catalysts have lower catalytic activity toward ethanol. Researchers have shown that ethanol oxidation at polycrystalline platinum surfaces showed carbon dioxide, acetaldehyde, and acetic acid as products [10], but at high concentration the major products are carbon dioxide and acetaldehyde [11–12]. This means that a portion of ethanol is completely oxidized to carbon dioxide (12 electron process) via the reaction above and a portion of ethanol is partially oxidized through the following 2-electron process:

$$CH_3CH_2OH \rightarrow CH_3CHO + 2H^+ + 2e^-$$

However, in the absence of water, the ethanol reacts with 2 ethanol molecules to form ethanol diethylacetal [13]. It is important to note that the efficiency of the system is quite different for ethanol than methanol. Methanol oxidation shows approximately 90% of products are carbon dioxide, whereas ethanol oxidation varies between 20 and 40% depending on the catalyst [13]. Even though methanol oxidation has higher conversion efficiency, the methanol by-product of methanol oxidation has much higher toxicity than ethanol (OSHA exposure limits are 1 ppm for methanol and 200 ppm for ethanol and the LD_{50} during inhalation for rats or mice: 203 mg/m^3 for methanol and 24,000 mg/m^3 for ethanol [13]).

PtRu CATALYSTS

Alloys of platinum and ruthenium have become common electrocatalysts for fuel cells, because it is believed that alloying ruthenium with platinum will help increase the carbon monoxide tolerance of the platinum catalysts. Alloys of platinum and ruthenium have also been used extensively for DMFC fuel cells, along with hydrogen/oxygen fuel cells that employ hydrogen gas formed from a reformation process that may have carbon monoxide or carbon monoxide-like by-products. Although

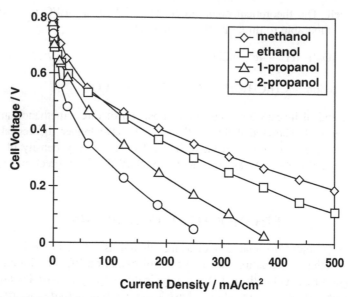

FIGURE 10.2 Comparison of fuel cell performance for 4 different alcohol fuels employing a 4-mg/cm² Pt/Ru catalyst at the anode and a 4-mg/cm² platinum black at the cathode. *Source*: Wang, J., Wasmus, S. and Savinell, R.F., *J. Electrochemical Society*, 142, 4218, 1995. With permission.

extensive research was done on Pt/Ru alloys on carbon supports and platinum on carbon supports, there was no statistical difference between the selectivity of the two catalysts for ethanol electrooxidation [13]. Figure 10.2 shows a comparison of fuel cell performance for different alcohol fuels employing Pt/Ru alloys as catalysts. It is apparent that methanol performance is better at high current densities (at a current density of 250mA/cm², the cell voltages are 0.354V for methanol, 0.305V for ethanol, 0.174V for 1-propanol, and 0.054V for 2-propanol [13]), but ethanol performance is better at low current densities (>0.05V at low current densities). The excellent performance of ethanol at low current density is likely due to a decrease in crossover of ethanol versus methanol to the cathode. It is also interesting to note that propanol performance is significantly worse than methanol and ethanol. 1-propanol oxidation forms carbon dioxide and propionaldehyde, but 2-propanol oxidation forms carbon dioxide and acetone [14]. The direct alcohol fuel cells studied in Figure 10.2 are being operated at a temperature of 170°C [13]. This temperature is extremely high (harsh enough that Nafion is not particularly stable and another polymer electrolyte (polybenzimidazole) was used) and is above the temperatures that are realistic for portable power applications, but they provide a benchmark for comparing the 4 alcohols.

PtSn Catalysts

Zhou and coworkers have studied the effect of other alloys on ethanol electrooxidation. Figure 10.3 shows representative cyclic voltammograms of alloys of

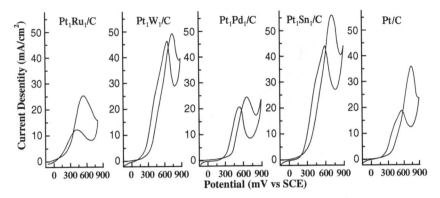

FIGURE 10.3 Representative cyclic voltammograms of ethanol oxidation at different anode catalysts in a solution of 1.0 M ethanol and 0.5 M sulfuric acid. The voltammograms were taken at a scan rate of 10 mV/s at 25°C. *Source*: Wang, J., Wasmus, S. and Savinell, R.F., *J. Electrochemical Society*, 142, 4218, 1995. With permission.

platinum with ruthenium, tungsten, palladium, and tin. These voltammograms were taken at room temperature in solutions that contain 1.0 M ethanol and 0.5 M sulfuric acid. The voltammograms show the largest catalytic activity (current density at the oxidation peak) for PtSn on carbon, but the PtRu on carbon has the lowest over-potential for the ethanol oxidation peaks (0.23 V lower than pure platinum on carbon) [15]. Figure 10.4 shows voltage-current curves and power curves for the same catalysts in a direct ethanol fuel cell at 90°C. The results indicate that Sn, Ru, and W increase the catalytic activity for ethanol oxidation on platinum (maximum power density of 52.0 mW/cm^2, 28.6 mW/cm^2, and 16.0mW/cm^2, respectively, compared to 10.8 mW/cm^2 for pure platinum on carbon [15]). Tin and ruthenium are believed to have a bifunctional mechanism to supply surface oxygen containing species for the oxidative removal of carbon monoxide like species that typically passivate the surface of pure platinum [16]. The proposed mechanism for ethanol oxidation at Pt/Sn alloys is shown below [17]:

$$C_2H_5OH + Pt(H_2O) \rightarrow Pt(C_2H_5OH) + H_2O$$

$$Pt(C_2H_5OH) + Pt \rightarrow Pt(CO) + Pt(res) + xH^+ + xe^-$$

$$Pt(C_2H_5OH) \rightarrow Pt(CH_3CHO) + 2H^+ + 2e^-$$

$$Pt(CH_3CHO) + SnCl_4(OH)_2^{2-} \rightarrow CH_3COOH + SnCl_4^{2-} + 2H_2O + e^- + H^+$$

$$Pt(CO), Pt(res) + SnCl_4(OH)_2^{2-} \rightarrow CO_2 + SnCl_4^{2-} + H_2O + Pt$$

$$H_2O + Pt = Pt(OH)_{ads} + e^- + H^+$$

$$2Pt(OH)_{ads} + SnCl_4^{2-} = SnCl_4(OH)_2^{2-} + 2Pt$$

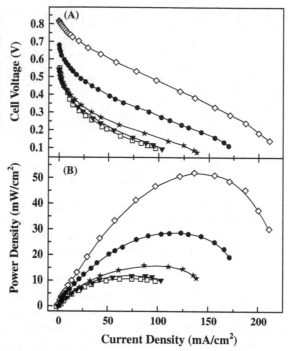

FIGURE 10.4 Comparison of voltage current curves and power curves for 1.0 M ethanol fuel cells at 90°C with different anode catalysts: □ – Pt/C (2.0 mg Pt/cm²), ▼ – PtPd/C (1.3 mg Pt/cm²), * – PtW/C (2.0 mg Pt/cm²), ● – PtRu/C (1.3 mg Pt/cm²), and ◊ PtSn/C (1.3 mg Pt/cm²). The ethanol fuel solution was pumped at 1.0 mL/min. The PEM was Nafion 115 and the cathode was a 20% Pt on Vulcan XC-72 carbon support with a loading of 1.0 mg Pt/cm². *Source*: Zhou, W.J., Li, W.Z., Song, S.Q., et al., *Power Sources*, 131, 217, 2004. With permission.

where Pt(res) is an oxidized residue adsorbed to the surface of platinum, Pt(H₂O) is water adsorbed to the surface of platinum, Pt(CO) is carbon monoxide adsorbed to the surface of platinum, Pt(C₂H₅OH) is ethanol adsorbed to the surface of platinum, and Pt(CH₃CHO) is acetaldehyde adsorbed to the surface of platinum.

After Pt/Sn alloys were determined to be the optimal elemental alloy, Zhou and coworkers examined the importance of tin content and temperature on the fuel cell power curves. Figure 10.5 shows the effect of altering the tin content on the direct ethanol fuel cell performance at a temperature of 60°C. The figure shows both current voltage curves and power curves. The results clearly show that Pt₃Sn₂ on carbon is the best catalyst choice for 60°C [18]. Figure 10.6 shows the effect of altering tin catalyst content on the fuel cell performance at a temperature of 90°C. The results clearly show that Pt₂Sn₁ on carbon is best for temperatures that are greater than 75°C [18]. Figure 10.5 and Figure 10.6 show that tin content does affect fuel cell performance and temperature affects the catalytic activity of each fuel cell differently. The operating temperature for DEFC

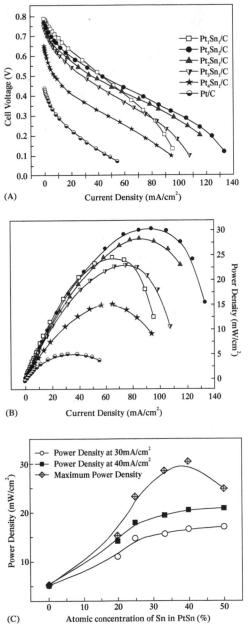

FIGURE 10.5 Fuel cell performance data for different tin catalyst content at 60°C. The anode catalyst notation corresponds to different Pt/Sn atomic ratios with a constant platinum loading of 1.3 mg/cm². The cathode catalyst has a loading of 1.0 mg/cm². Both cathode and anode catalysts are supported on Vulcan XC-72 carbon. The PEM is Nafion 115 and the cell is run in 1 M ethanol at a flow rate of 1 mL/min. *Source*: Zhou, W.J., Song, S.Q., Li, W.Z., et al., *Power Sources*, 140, 50, 2005. With permission.

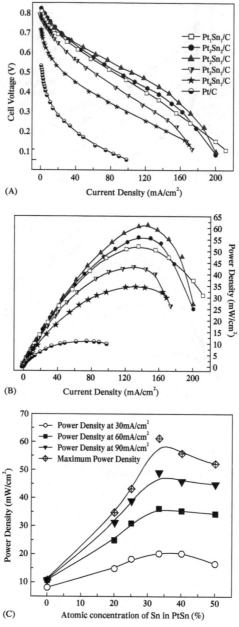

FIGURE 10.6 Fuel cell performance data for different tin catalyst content at 90°C. The anode catalyst notation corresponds to different Pt/Sn atomic ratios with a constant platinum loading of 1.3 mg/cm². The cathode catalyst has a loading of 1.0 mg/cm². Both cathode and anode catalysts are supported on Vulcan XC-72 carbon. The PEM is Nafion 115 and the cell is run in 1 M ethanol at a flow rate of 1 mL/min. *Source*: Zhou, W.J., Song, S.Q., Li, W.Z., et al., *Power Sources*, 140, 50, 2005. With permission.

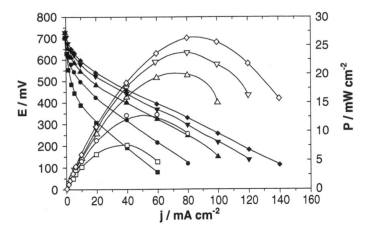

FIGURE 10.7 Fuel cell performance as a function of temperature for a Pt:Sn (9:1)/C anode where □ – 50C, ○– 70C, △– 90°C, ▽– 100°C, and ◊– 110°C. Catalyst loading was 1.5 mg/cm² and was dispersed on Vulcan XC-72 at a loading of 30% by wt. catalyst. The cathode was an E-Tek 40% Pt/XC72 cathode with 2.0 mg/cm² platinum catalyst loading. The PEM was a Nafion 117 membrane. *Source*: Lamy, C., Rousseau, S., Belgsir, E.M., et al., *Electrochimica Acta*, 49, 3901, 2004. With permission.

is a function of application. Most portable power applications need to operate between room temperature and 50°C, but performance tends to increase with temperature until crossover and/or polymer electrolyte membrane degradation take over. At 30°C, DEFC have maximum power densities that range from 2 to 10 mW/cm² [19]. Figure 10.7 shows the effect of a wider range of temperatures (50°C–110°C) for a DEFC with a Pt-Sn (9:1)/C anode. It is important to note that fuel cell performance is a function of temperature and a degradation is not seen at high temperatures [5]. Open circuit potentials do not vary significantly with temperature, but maximum power ranges from 6 to 26 mW/cm² [5].

Catalyst loading and catalyst supports have also been investigated as parameters that may affect DEFC performance. Studies in hydrogen/oxygen and DMFC have shown that loading of the catalyst can affect fuel cell performance. If the catalyst loading of the DEFC in Figure 10.5 is changed from 30% metal on vulcanized carbon XC-72 to 60% metal on vulcanized carbon XC-72, the maximum power can increase to 28 mW/cm² and the open circuit potential can increase from 0.72V to 0.75 V [5]. Research has also shown that transitioning from vulcanized carbon supports (XC-72) to multiwall carbon nanotubes (MWNTs) increases both the open circuit potential and the maximum power density of a DEFC with a platinum/tin alloy catalyst [9]. This is shown in Figure 10.8 where the open power curve shows an increase from 30 mW/cm² to 38 mW/cm² with an increase of 80 mV in open circuit potential.

DEFCs can be fabricated by methods similar to DMFCs. The most common format is the membrane electrode assembly (MEA). A MEA is a single assembly that contains the anode, the cathode, and the polymer electrolyte membrane

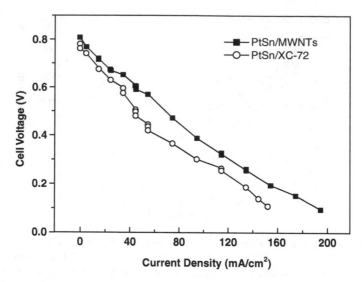

FIGURE 10.8 Voltage current curve for two DEFC with the same platinum/tin alloy on different carbon substrates at 75°C and a concentration of 1 M ethanol. Anode and cathode loading was 1.0 mg/cm^2 platinum. Nafion 115 was used as the polymer electrode membrane and the flow rate was 1mL/min. *Source*: Zhao, X., Li, W., Jiang, L., et al., *Carbon*, 42, 3251, 2004. With permission.

in ionic contact with each other. MEAs are formed most commonly using a conventional heat pressing method, but they can also be fabricated using a decal transfer method. The conventional method involves sandwiching the PEM between an anode and cathode and heat pressing the sandwich at a temperature above the glass transition temperature of the PEM to melt the electrodes into ionic contact with the PEM. The conventional method shows a 34% decrease in power density over a 10-hour period and delamination of the electrodes from the Nafion [20]. This is likely due to the increased swelling of Nafion in the presence of ethanol, but the decal transfer method only shows a 15% decrease and no delamination, along with no change in resistance [20]. Therefore, the decal transfer method is a better method for forming DEFC MEAs. The decal transfer method involves spray painting the catalyst layer directly onto the polymer electrolyte membrane, instead of onto an electrode support (such as carbon paper) and then heat pressing into the polymer electrolyte membrane.

Researchers have also studied tertiary catalyst systems, but the fuel cell performance has not been greatly affected by adding a third component to the system for alloys containing platinum and ruthenium with a third component of tungsten, tin, or molybdenum [15]. Tertiary catalysts with tungsten and tin did show a measurable increase in power compared to pure Pt/Ru alloys, but both power densities are less than pure Pt/Sn alloys under the same operating conditions [15].

CONCLUSIONS

Direct ethanol fuel cells are a relatively new technology for portable power generation. Results have concluded that electrochemical oxidation of ethanol on platinum-based catalysts is not significantly lower than for methanol [21] and the intermediate products of ethanol oxidation are less toxic than methanol oxidation. Although catalytic performance with pure platinum catalysts is low, the performance of Pt/Sn and Pt/Ru alloys is good. Future research will focus on the development of electrocatalysts that show improved catalytic activity and lower electrode fouling at low and moderate temperatures (room temperature to 50°C). Research on fuel cell lifetimes will also be conducted to study the long-term effects of continuous operation on the catalysts, electrode support, and polymer electrolyte membranes. Improved lifetimes are an issue for both methanol and ethanol, because both oxidation processes produce carbon monoxide and carbon monoxide-like products that adsorb/passivate the catalyst and both alcohols swell the polymer electrolyte membrane, which typically decreases the lifetime and stability of the membrane. Overall, DEFCs are a relatively new technology compared to DMFCs, but ethanol has advantages over methanol in decreased toxicity and environmental issues. From a catalytic perspective, the catalytic rates are similar between ethanol and methanol oxidation, but methanol oxidation is more efficient (produces a larges percentage of carbon dioxide (the complete oxidation product)).

REFERENCES

1. Ren, X., Wilson, M. and Gottesfeld, S., *J. Electrochem. Soc.*, 143, 12, 1996.
2. Ren, X., Zelenay, P., Thomas, S., Davey, J. and Gottesfeld, S., *J. Power Sources*, 86, 111, 2000.
3. Thomas, S.C., Ren, X.M., Gottesfeld, S. and Zelenay, P., *Electrochimica Acta*, 47, 3741, 2002.
4. Glaebrook, W., *J. Power Sources*, 7, 215, 1982.
5. Lamy, C., Rousseau, S., Belgsir, E.M., Coutanceau, C. and Leger, J.M., *Electrochimica Acta*, 49, 3901, 2004.
6. *Energy from Biological Processes: Technical and Policy Opinions*, Office of Technology Assessment, Westview Press, 1981.
7. Lamy, C., Belgsir, E.M. and Leger, J.M., *J. Applied Electrochemistry*, 31, 799, 2001.
8. Lamy, C., Lima, A., LeRhun, V., Delime, F., Cutanceau, C. and Leger, J.M., *J. Power Sources*, 105, 283, 2002.
9. Zhao, X., Li, W., Jiang, L., Zhou, W., Xin, Q., Yi, B. and Sun, G., *Carbon*, 42, 3251, 2004.
10. Gao, P., Chang, S., Zhou, Z. and Waver, M.J., *J. Electroanal. Chem.*, 272, 161, 1989.
11. Willsau, J. and Heitbaum, J., *J. Electrochtroanal. Chem.*, 194, 27, 1985.
12. Bittins-Cattaneo, B., Wilhelm, S., Cattaneo, E., Buschmann, H.W. and Vielstich, W. *Ber. Bunsenges. Phys. Chem.*, 92, 1210, 1988.

13. Wang, J., Wasmus, S. and Savinell, R.F., *J. Electrochemical Society* 142, 4218, 1995.

14. Hartung, T., Ph.D. thesis, University of Witten, Germany, 1988.

15. Zhou, W.J., Li, W.Z., Song, S.Q., Zhou, Z.H., Jiang, L.H., Sun, G.Q., Xin, Q., Poulianitis, K., Kontou, S. and Tsiakaras, P., *J. Power Sources*, 131, 217, 2004.

16. Gurau, B., Viswanathan, R., Liu, R. and Lafrenz, T.J., *J. Phys. Chem. B*, 102, 9997, 1998.

17. Jiang, L., Zhou, Z., Li, W., Zhou, W., Song, S., Li, H., Sun, G. and Xin, Q., *Energy and Fuels*, 18, 866, 2004.

18. Zhou, W.J., Song, S.Q., Li, W.Z., Zhou, Z.H., Sun, G.Q., Xin, Q., Douvartzides, S. and Tsiadkaras, P., *J. Power Sources*, 140, 50, 2005.

19. Jiang, L., Zhou, Z., Wang, S., Liu, J., Zhao, X., Sun, G., Xin, Q. and Zhou, B., *Preprints of the American Chemical Society Division of Fuel Chemistry* 49, 668, 2004.

20. Song, S., Wang, G., Zhou, W., Zhao, X., Sun, G., Xin, Q., Kontou, S., and Tsiakaras, P., *J. Power Sources*, 140, 103, 2005.

21. Arico, A.S., Creti, P.L. and Antonucci, P.L., *Electrochemical and Solid-State Letters*, 1, 66, 1998.

11 Solid-Oxide Fuel Cells Operating with Direct-Alcohol and Hydrocarbon Fuels

Fatih Dogan
Department of Materials Science and Engineering
University of Missouri-Rolla

CONTENTS

Abstract This chapter addresses utilization of alcohol and other hydrocarbon-based fuels to generate electricity in solid-oxide fuel cells (SOFCs). One of the key advantages of SOFC is that both external as well internal fuel reforming is possible to operate the fuel cell under stable conditions. While alcohol fuels can be obtained sulfur-free and in high purity, hydrocarbon fuels have higher energy density and existing infrastructure of production and distribution. Development of more energy-efficient and chemically stable electrode materials is necessary for SOFC operating at high (800–1000°C) and intermediate (500–800°C) temperatures. Significant progress has been made in recent years in the development of carbon monoxide-tolerant fuel electrodes (anodes) to prevent carbon deposition on the catalyst that results in a reduced performance of the fuel cell. Development of fuel electrodes compatible with alcohol and hydrocarbon fuels will lead to more efficient and widespread applications of SOFCs in double-chamber and single-chamber modes.

INTRODUCTION

Fuel cells are viewed as environmentally compatible and efficient energy conversion systems. A fuel cell works much like a battery with external fuel supplies. Chemical fuels are electrochemically converted into electricity at high efficiencies without producing significant amount of pollutants such as nitrogen oxides as compared to combustion engines. Hydrogen is the ideal fuel since it reacts with oxygen in the air to produce water and an electric current, but hydrogen is expensive and difficult to store. Until the hydrogen economy is well established, other fuels can be used indirectly with an external reformer or directly to operate fuel cells. Hydrogen is stored naturally in alcohols (e.g., ethanol and methanol) or hydrocarbons such as propane and methane, which are available to produce cleaner power if the electrochemical processes of hydrocarbon oxidation reactions are well understood.

Among various fuel cells, solid oxide fuel cells (SOFCs) and molten carbonate fuel cells can be operated using hydrogen as well as carbon monoxide. Particularly, SOFC is viewed as the most flexible fuel cell system that can operate using various fuel gases directly supplied to the fuel electrodes [1–3]. Removal of CO from H_2 fuel is essential for polymer electrolyte membrane fuel cells, which are generally considered to be the most viable approach for mobile applications.

The application of high and intermediate temperature SOFCs range from small-scale domestic heat and power to large-scale distributed power generation. SOFCs offer high efficiencies up to 60–70% in individual systems and up to 80% in hybrid systems by extracting the energy present in the high-temperature exhaust gases, e.g., by using gas or steam turbines [4]. High-temperature SOFC applications include multimegawatt-scale centralized power generation, distributed power generation up to 1 MW and combined heat/power (CHP) plants in the 100-kW to 1-MW range. Potential areas of application for intermediate SOFCs are in the transport sector (up to 50 kW), military and aerospace (5 to 50 kW), domestic CHP (up to 10 kW) and miniaturized fuel cells "palm-power" in the 10-W range.

In SOFC, the electrolyte is typically a dense yttria-stabilized zirconia (YSZ), which is an ionic conductor blocking electron transport as shown in Figure 11.1. The electrolyte allows the transport of oxygen ions via the oxygen vacancies from the interface at the air electrode (cathode) to the interface with the fuel electrode (anode). The cathode is typically composed of a porous lanthanum strontium manganese oxide with YSZ and facilitates the reaction for the reduction of oxygen gas to oxygen ions at the electrode/electrolyte interface. The anode material is typically a porous Ni-YSZ composite allowing the oxidation of the fuel and transport of the electrons from the electrolyte/electrode interface to the interconnect of the fuel cell stack. The interconnect material is typically lanthanum strontium chromite for high-temperature operation while corrosion-resistant metallic alloys are employed in the development of SOFCs operating at intermediate temperatures. The role of the interconnect is to transfer

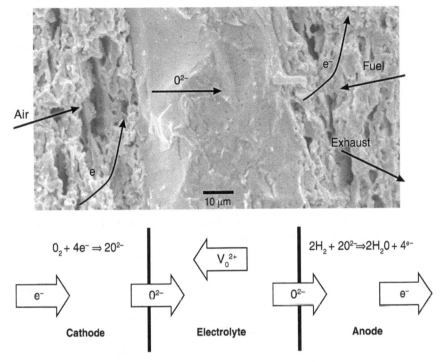

FIGURE 11.1 The microstructure of a typical SOFC and the electrochemical reactions occurring at the interface between the electrodes and electrolyte. Reprinted with permission from [5]. Nature Publishing Group.

the electrons between the individual cells in the stack and to prevent mixing of fuel and oxidant gases [5].

A diverse range of fuels can be used in SOFCs since the internal temperature is high enough to initiate fuel conversion reactions. Hence, SOFCs have an efficiency advantage over polymer electrolyte membrane fuel cells when alcohol or hydrocarbon fuels are to be used, even though direct-methanol fuel cells with polymer electrolyte membranes are widely studied. The use of these fuels in SOFCs without preprocessing, however, requires further advances in development of appropriate electrode materials toward preventing unwanted reactions such as carbon formation on the anode, which significantly affects the performance of the fuel cell.

FUELS FOR SOLID-OXIDE FUEL CELLS

Until the hydrogen economy is well established, it is more sensible to generate electricity directly from alcohols or hydrocarbons. SOFCs may become very attractive for portable, transportation, and stationary applications if alcohols and hydrocarbons can be utilized directly without applying any fuel pretreatments. The main advantage of liquid hydrocarbons is their relatively higher energy density compared to alcohols. However, most hydrocarbon fuels such as natural gas, bioderived gases, diesel, and gasoline contain impurities such as hydrogen sulfide and halogens, which may lead to poisoning of the SOFC electrode materials. Particularly, sulfur content in such fuels should be reduced through pretreatments to prevent the fuel cell electrodes from poisoning. Alternatively, progress is being made toward development of sulfur-resistant electrode materials for long-lasting operation of SOFCs using hydrocarbon fuels, which generally contain sulfide compounds in relatively high concentrations. For instance, a highly sulfur-tolerant anode composed of Cu, CeO_2, and YSZ was developed to operate a SOFC using hydrogen with H_2S levels up to 450 ppm at 1073 K [6]. Another study based on $La_xSr_{1-x}VO_{3-}$ as anode material for SOFC showed a maximum power density of 135 mW/cm² at 280 mA/cm² when the fuel was a 5% H_2S–95% H_2 mixture at 1273K [7].

The advantage of liquid oxygenated hydrocarbons, such as alcohols, in comparison to gasoline is that they are cleaner (low sulfur content) and can be derived from agricultural by-products and biomass as a renewable energy source. Alcohol is an ideal fuel for the fuel cells because of ease of transportation, storage, and handling, as well as their high energy density. Partially oxidized (hydrated) fuels may be easily reformed, such as alcohols, as they contain oxygen, in a liquid form. Since water is often used for internal reforming of the fuel, water solubility of alcohols (especially methanol, ethanol, and propanol) offers the advantage that additional fuel processing may not be necessary for operation of the fuel cell.

An anode-supported SOFC utilizing direct alcohol was reported by Jiang and Virkar [8]. A thin-film YSZ electrolyte was deposited on a Ni–YSZ anode with a composite of Sr-doped $LaMnO_3$ and YSZ as a cathode. Pure methanol and an equivolume mixture of ethanol and water were used as fuels to operate the cells over a range of temperatures. Power densities achieved with ethanol and water mixtures were between 0.3 W/cm² at 650°C and 0.8 W/cm² at 800°C, and with methanol between 0.6 W/cm² at 650°C and 1.3 W/cm² at 800°C as shown in Figures 11.2 and 11.3. Carbon deposition on the electrodes was not observed when methanol was used as fuel. On the other hand, maximum power density using humidified H_2 was 1.7 W/cm² at 800°C. This indicates that a lack of H_2 in the fuel may substantially increase concentration polarization thus limiting the performance of the cell.

Another study on direct-alcohol SOFCs reported a comparison of methanol, ethanol, propanol, and butanol as fuel sources [9]. With an increasing carbon number of the alcohol, a decrease in cell voltage was observed, which was attributed to slower decomposition and/or reforming kinetics of alcohols. Decreasing operational temperatures led to an increase of unreacted alcohols, aldehydes, and aromatic compounds. Thermochemical calculations were used to reveal the equilibrium amounts of reaction products of fuels during fuel cell operation [10,11]. Figure 11.4 shows the limit lines of carbon deposition as a function of temperature in the C–H–O diagram. No carbon deposition is expected if the carbon-to-oxygen ratio is less than unity. It is shown that the addition of

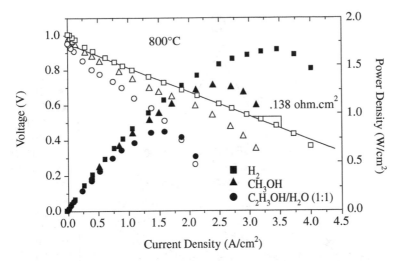

FIGURE 11.2 Cell performance at 800°C with methanol, ethanol, and hydrogen as fuels. Flow rate for hydrogen was 100 mL/min. Methanol used was undiluted. Ethanol used was an equivolume solution of ethanol and water. The flow rates for liquid fuels were 0.2 mL/min for both. *Source*: Jiang, Y. and Virkar, A., High Performance, Anode-Supported Solid Oxide Fuel Cell Operating on Hydrogen Sulfide (H_2S) and Sulfur-Containing Fuels, *J. Power Sources*, 2004. With permission. Copyright [2001], The Electrochemical Society.

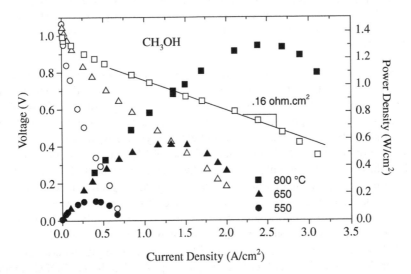

FIGURE 11.3 Cell performance with methanol as a fuel at various temperatures. The flow rate was 0.2 mL/min. *Source*: Jiang, Y. and Virkar, A., High Performance, Anode-Supported Solid Oxide Fuel Cell Operating on Hydrogen Sulfide (H_2S) and Sulfur-Containing Fuels, *J. Power Sources*, 2004. With permission. Copyright [2001], The Electrochemical Society.

H_2O, O_2 and/or CO_2 is necessary to prevent the carbon deposition since the positions of various fuels are within the deposition region.

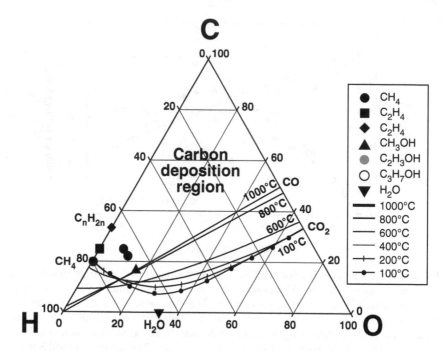

FIGURE 11.4 Carbon deposition limit lines for various fuels and temperatures in the C–H–O diagram. Reprinted with permission from [11]. Copyright [2003], The Electrochemical Society. *Source*: Sasaki, K. and Teraoka, Y., *J. Electrochem. Soc.*, 150(7), 2003. With permission.

Further development of electrode materials that do not require introduction of water, will lead to better performance of the SOFCs, provided carbon deposition can be suppressed. Recent studies showed that coking issues can be resolved through selection of appropriate catalysts and anode materials in fuel cell development [4,5]. Because nickel is an excellent catalyst for hydrocarbon cracking, Ni/ZrO_2 cermets are used as anode materials for YSZ-based SOFCs. As mentioned earlier, these cermets can only be used in hydrocarbon or alcohol fuels if excess water is present to ensure complete fuel reforming. Mixing *iso*-octane with water, alcohol, and surfactant to produce an oil in water microemulsion was successful in reducing the carbon formation significantly, while retaining a high octane number [12]. It has been shown that the problem of carbon deposition may be avoided by using a copper–ceria anode [13] or applying an yttria–ceria interface between YSZ and Ni–YSZ cermet anode [3]. A nickel-free SOFC anode, $La_{0.75}Sr_{0.25}Cr_{0.5}Mn_{0.5}O_3$ with comparable electrochemical performance to Ni/YSZ cermets was developed for methane oxidation without using excess steam [5]. A recent study showed that a $Ru–CeO_2$ catalyst layer with a conventional anode allows internal reforming of *iso*-octane without cocking and yields stable power densities of 0.3 to 0.6 W/cm^2 in a SOFC design operating at intermediate temperatures [14].

SINGLE-CHAMBER SOLID-OXIDE FUEL CELLS AND HYDROCARBON FUELS

A single-chamber solid-oxide fuel cell (SC-SOFC), which operates using a mixture of fuel and oxidant gases, provides several advantages over the conventional double-chamber SOFC, such as simplified cell structure (no sealing required) and direct use of hydrocarbon fuel [15,16]. Figure 11.5 shows a schematic diagram of SC–SOFC operation. The oxygen activity at the electrodes of the SC–SOFC is not fixed and one electrode (anode) has a higher electrocatalytic activity for the oxidation of the fuel than the other (cathode). Oxidation reactions of a hydrocarbon fuel can be represented with a simplified multistep, quasi-general mechanism as follows:

$$C_nH_m + (n/2)O_2 \rightarrow nCO + (m/2)H_2 \qquad (11.1)$$

$$H_2 + O^{2-} \rightarrow H_2O + 2e^- \qquad (11.2)$$

$$CO + O^{2-} \rightarrow CO_2 + 2e^- \qquad (11.3)$$

On the other hand, the cathode has a higher electrocatalytic activity for the reduction of oxygen according to the reaction:

$$1/2O_2 + 2e^- \rightarrow O^{2-} \qquad (11.4)$$

These reactions lead to a low oxygen partial pressure at the anode locally, while the oxygen partial pressure at the cathode remains relatively high. As a

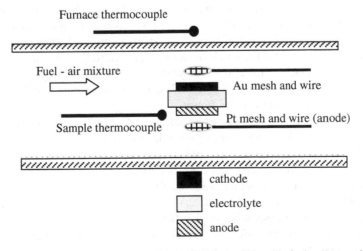

FIGURE 11.5 Schematic diagram of a single-chamber solid-oxide fuel cell operating with a mixture of fuel and air.

result, an electromotive force (emf) between two electrodes is generated with a mixed fuel and air mixture. Due to the presence of oxygen at the anode, SC–SOFC is not affected by the problems associated with carbon deposition, which is a significant drawback for double-chamber SOFCs when Ni–cermet is used as anode material.

The fuel/air mixtures for SC–SOFC were generally chosen to be richer than the upper explosion limits, yet they were fuel-lean enough to prevent the carbon deposition, which has been a significant problem in double-chamber SOFCs [17]. However, variations in the ratios of the local fuel-air mixture were also dependent on catalytic activity and test conditions that affect the performance of the fuel cell [15]. An ideal SC–SOFC has the same open circuit voltage (OCV) and I-V output as a double-chamber cell, given a uniform oxygen partial pressure. The difference in catalytic properties of the electrodes must be sufficient to cause a significant difference in oxygen partial pressure between the anode and the cathode. For the ideal SC–SOFC, one electrode would be reversible toward oxygen adsorption and inert to fuel, while the other electrode would be reversible toward fuel adsorption and completely inert to oxygen [18]. If the electrode materials are not sufficiently selective, a parasitic reaction creates mixed potentials at the electrodes, which reduces the efficiency of the cell. Compared to traditional double-chamber fuel cells, parasitic reactions in a single-chamber fuel cell have historically reduced the OCV by about half. This is analogous to a leak that allows the fuel to seep into the oxidizer side of a double-chamber fuel cell [19]. Advances in electrode catalyst materials for SC–SOFC may lead to a similar performance as a conventional double-chamber SOFC with a substantial reduction in complexity and cost of the fuel cell.

Significant improvement in the performance of single-chamber solid-oxide fuel cells has been achieved in recent years [15, 20–22]. Since SC–SOFC does not require high-temperature sealing materials to prevent the mixing of fuel gas and oxygen at operation temperatures, it offers a robust and more reliable alternative to double-chamber SOFC for special applications. As further advances are made toward controlling the catalytic activity of electrode materials, electrolyte resistance particularly at lower operating temperatures, optimizing of the gas flow rate and the cell configuration, SC–SOFCs may find widespread implementation as compact power sources in the future.

Several recent studies on the development of SC–SOFCs have been conducted in our laboratory to improve their performance and understand complex electrode reactions [20,23–26]. Initial experiments were carried out using fuel cells prepared by deposition of YSZ thin-film electrolytes (1-2 μm thickness) on the NiO–YSZ anode as a substrate with (La,Sr)(Co,Fe)O_3 (LSCF) as the cathode (Figure 11.6). A power density of 0.12 W cm^{-2} was obtained at an OCV of >0.8 V using a methane-air gas mixture as a fuel [23].

In another study, the effect of mixed gas flow rates on the performance of SC–SOFCs has been investigated using a cell that consists of a 18-μm thick YSZ porous electrolyte on a NiO–YSZ anode substrate with a (La, Sr)(Co, Fe)O_3 cathode. Higher gas flow rates led to an increase of cell temperature due to

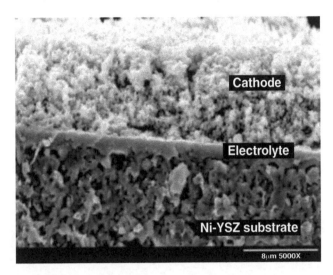

FIGURE 11.6 Microstructural development of a fuel cell with a dense YSZ electrolyte (1–2µm thick) prepared by a low-temperature processing method (annealed at 950°C). *Source*: Suzuki, T., Jasinski, P., Anderson, H., et al., *J. Electrochem. Soc.*, 151(9), 2004. With permission.

increasing methane reaction rate, which resulted in improved cell performance. Figure 11.7 shows that optimization of gas flow rate (linear velocity) lead to a decrease of the operating temperature effectively and increased cell performance as well as fuel efficiency. At a cell temperature of 744°C (furnace temperature: 606°C), an OCV of ~0.78 V and a maximum power density of ~660 mW cm^{-2} (0.44 V) were obtained. The results indicated that a porous ion-conducting membrane provides sufficient separation of oxygen activity at the electrodes by selection of an optimum operation temperature and a gas flow rate. Thus, it appears that SC–SOFCs with porous electrolyte provide opportunities to design thermally and mechanically more robust stacks by utilizing hydrocarbon fuels. It also allows fabrication of the cells at lower temperatures using conventional processing techniques such as screen printing, since densification of the electrolyte at high sintering temperatures is not required.

SUMMARY

Solid oxide fuel cells are very promising energy conversion systems that can generate electricity at high efficiencies using not only hydrogen but also alcohol and hydrocarbon fuels. Further progress on the development of fuel cell materials, particularly the electrodes, which prevents carbon deposition and sulfur-resistance, will play a key role to achieve a stable operation of direct-alcohol SOFCs with high power densities. In contrast to double-chamber SOFCs, the single-chamber solid-oxide fuel cells offer a simple design that are not affected by the challenges of high temperature sealing and may be a cost-effective alternative

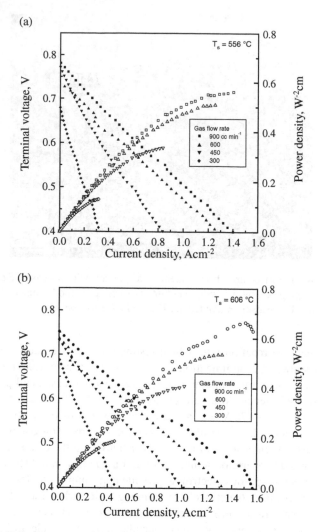

FIGURE 11.7 Performance of SC–SOFC with a porous YSZ electrolyte as a function of furnace set temperature and methane-air gas flow rate. *Source*: Suzuki, T., Jasinski, P., Petrovsky, V., et al., *J. Electrochem. Soc.*, 152(3), 2005. With permission.

with a mechanically more robust structure. Recent developments in single-chamber SOFCs show promising results toward achieving significantly high power densities using hydrocarbon fuel gases mixed with air.

REFERENCES

1. Steele, B.C.H., Fuel-Cell Technology—Running on Natural Gas, *Nature*, 400, 619–621, 1999.

2. Putna, E.S., Stubenrauch, J., Vohs, J.M. and Gorte, R.J., Ceria-Based Anodes for the Direct Oxidation of Methane in Solid Oxide Fuel Cells, *Langmuir*, 11 (12), 4832–4837, 1995.

3. Murray, E.P., Tsai, T. and Barnett, S.A., A Direct-Methane Fuel Cell with a Ceria-Based Anode, *Nature*, 400, 649–651, 1999.

4. Song, C., Fuel Processing for Low-Temperature and High-Temperature Fuel Cells: Challenges, and Opportunities for Sustainable Development in the 21st Century, *Catalysis Today*, 77, 17–49, 2002.

5. Atkinson, A., Barnett, S., Gorte, R.J., Irvine, J.T.S., McEvoy, A.J., Mogensen, M., Singhal, S.C. and Vohs, J., Advanced Anodes for High-Temperature Fuel Cells, *Nature Materials*, 3, 17–27, 2004.

6. He, H.P., Gorte, R.J. and Vohs, J.M., Highly Sulfur Tolerant Cu-Ceria Anodes for SOFCs, *Electrochem. Solid State Lett.*, 8 (6), A279, A280, 2005.

7. Aguilar, L., Zha, S.W., Cheng Z., Winnick, J. and Liu, M.L., A Solid Oxide Fuel Cell Operating on Hydrogen Sulfide (H_2S) and Sulfur-Containing Fuels, *J. Power Sources*, 135 (1-2), 17–24, 2004.

8. Jiang, Y. and Virkar, V.A., High Performance, Anode-Supported Solid Oxide Fuel Cell Operating on Direct Alcohol, *J. Electrochem. Soc.*, 148 (7), A706–A709, 2001.

9. Sasaki, K., Watanabe, K. and Teraoka, Y., Direct-Alcohol SOFCs: Current-Voltage Characteristics and Fuel Gas Compositions, *J. Electrochem. Soc.*, 151 (7), A965–A970.

10. Sasaki, K. and Teraoka, Y., Equilibria in Fuel Cell Gases—I. Equilibrium Compositions and Reforming Conditions, *J. Electrochem. Soc.*, 150 (7), A878–A884, 2003.

11. Sasaki, K. and Teraoka, Y., Equilibria in Fuel Cell Gases—II. The C-H-O Ternary Diagrams, *J. Electrochem. Soc.*, 150 (7), A885–A888, 2003.

12. Saunders, G. J., Preece, J. and Kendall, K., Formulating Liquid Hydrocarbon Fuels for SOFCs, *J. Power Sources*, 131, 23–26, 2004.

13. Park, S., Vohs, J.M. and Gorte, R.J., Direct Oxidation of Hydrocarbons in a Solid-Oxide Fuel Cell, *Nature*, 404, 265–267, 2000.

14. Zhan, Z. and Barnett, S.A., An Octane-Fueled Solid Oxide Fuel Cell, 308, 844–847, 2005.

15. Hibino, T., Hashimoto, A., Inoue, T., Tokuno, S. and Sano, M.A., Low-Operating-Temperature Solid Oxide Fuel Cell in Hydrocarbon-Air Mixtures, *Science*, 288, 2031–2033, 2000.

16. Hibino, T., Hashimoto, A., Yano, M., Suzuki, M., Yoshida, S. and Sano, M.A., Solid Oxide Fuel Cell Using an Exothermic Reaction as the Heat Source, *J. Electrochem. Soc.*, 148, A544–A549, 2001.

17. Steele, B.C.H., Survey of Materials Selection for Ceramic Fuel Cells II. Cathodes and Anodes, *Solid State Ionics*, 1223, 86–88, 1996.

18. Riess, I., Vanderput, P.J. and Schoonman, J., Solid Oxide Fuel Cells Operating on Uniform Mixtures of Fuel and Air, *Solid State Ionics*, 82, 1–4, 1995.

19. Priestnall, M., Kotzeva, V., Fish, D. and Nilsson, E., Compact Mixed Reactant Fuel Cells, *J. Power Sources*, 106, 21–30, 2002.

20. Suzuki, T., Jasinski, P., Petrovsky, V., Anderson, H. and Dogan, F., Performance of a Porous Electrolyte in Single-Chamber SOFCs, *J. Electrochem. Soc.*, 152 (3), A527–A531, 2005.

21. Shao, Z.P., Haile, S.M., Ahn, J., Ronney, P.D., Zhan, Z.L. and Barnett, S.A., A Thermally Self-Sustained Micro Solid-Oxide Fuel-Cell Stack with High Power Density, *Nature*, 435, 795–798, 2005.

22. Stefan, I.C., Jacobson, C.P., Visco, S.J. and De Jonghe, L.C., Single Chamber Fuel Cells: Flow, Geometry, Rate, and Composition Considerations, *Electrochem. Solid-State Lett.*, 7, A198–A200, 2004.

23. Suzuki, T., Jasinski, P., Petrovsky, V., Anderson, H. and Dogan, F., Anode Supported Single Chamber Solid Oxide Fuel Cell in CH_4-Air Mixture, *J. Electrochem. Soc.*, 151 (9), A1473–A1476, 2004.

24. Suzuki, T., Jasinski, P., Anderson, H. and Dogan, F., Single Chamber Electrolyte Supported SOFC Module, *Electrochem. Solid-State Lett.*, 7 (11), A391–A393, 2004.

25. Jasinski, P., Suzuki, T., Dogan, F. and Anderson, H., Impedance Spectroscopy of Single Chamber SOFC, *Solid State Ionics*, 175 (1–4), 35–38, 2004.

26. Suzuki, T., Jasinski, P., Anderson, H. and Dogan, F., Role of Composite Cathodes in Single Chamber SOFC, *J. Electrochem. Soc.*, 151 (10), A1678–A1682, 2004.

12 Alcohol-Based Biofuel Cells

Sabina Topcagic, Becky L. Treu, and Shelley D. Minteer
Department of Chemistry, Saint Louis University, Missouri

CONTENTS

Abstract There are three types of batteries: primary, secondary, and fuel cells. A fuel cell is an electrochemical device that converts chemical energy into electrical energy via catalysts. Fuel cells have many advantages over the two other types of batteries due to the fact they can be regenerated with the addition of fuel specific to the system. Traditional fuel cells employ heavy metal or precious metal catalysts, whereas biofuel cells employ biological catalysts (enzymes). Enzymes are highly specific catalysts, so they allow for the simplification of the fuel cell by eliminating the need for a polymer electrolyte membrane, which is one of the mostly costly parts of a fuel cell. Dehydrogenase enzymes have been employed at the anode of biofuel cells to oxidize alcohols. Methanol, ethanol, propanol, and butanol are examples of alcohols that can be used in biofuel cells. Long-term goals include investigating a variety of power applications for this technology ranging from portable electronics to sensors.

INTRODUCTION

Previous chapters of this book detail methods for producing ethanol from agricultural products and biomass. Although many of these methods are efficient, it is crucial to be able to efficiently convert energy to electrical power. As detailed in an earlier chapter, researchers have been attempting to develop direct ethanol fuel cells (DEFCs), but there have been problems because traditional precious metal catalysts (Pt-based catalysts) are unable to efficiently catalyze the oxidation ethanol and maintain an electrode with minimal fouling at low temperatures. However, living organisms are capable of efficiently catalyzing the oxidation of ethanol at 20–40°C. Living organisms, such as *Pseudomonas aeruginosa* [1], acetobacter [2], and gluconobacter [3], contain enzymes that can oxidize a variety of alcohols, including ethanol. Over the last 40 years, researchers have been working on employing living organisms and/or their enzymes in a fuel cell to convert chemical energy to electrical energy. This type of battery or fuel cell is referred to as a biofuel cell. The early research was plagued with enzyme stability problems and low power densities, but those issues are being overcome with current research. In this chapter, we will discuss the brief history, techniques, and applications of alcohol-based biofuel cells.

PORTABLE ELECTRICAL ENERGY SOURCES

BATTERIES

Batteries are usually categorized as primary batteries, secondary batteries, or fuel cells [4]. Primary batteries cannot be recharged, because they have irreversible electrochemistry. They are single use and disposable. Examples of a primary battery include alkaline batteries (such as silver oxide/zinc, mercury oxide/zinc, and manganese oxide/zinc) [5]. Secondary batteries experience reversible electrochemistry, so they are reusable and can be recharged by an external power supply after the operating voltage has dropped to zero. Examples of a secondary battery include nickel-cadmium, nickel-metal-hydride, and lithium-ion batteries. Problems associated with secondary batteries are that they undergo hysteresis, which prohibits them from being recharged to their original state once used [4]. Unlike secondary batteries, fuel cells do not undergo hysteresis. A fuel cell is an electrochemical device that generates power upon fuel addition; therefore, it is not of single use nor does it need to be recharged by an external power source, but only by the addition of more fuel.

A battery is a portable, self-contained electrochemical power source that consists of one or more voltaic cells [5]. Single voltaic cells consist of two electrodes (an anode and a cathode) and at least one electrolyte. In all electrochemical power sources, electrodes are used to donate and accept electrons in order to generate power. Oxidation of fuel occurs at the anode electrode, while reduction occurs at the cathode electrode. Traditional electrode materials utilized in batteries are metal-based, such as platinum, nickel, lead, and lithium. Employment of these catalysts is limited due to the fact that they are nonrenewable resources and highly expensive. In addition, precious metal catalysts when employed at electrodes will oxidize a variety of fuels (hydrogen gas, methane, and alcohols: methanol, ethanol, propanol, butanol, other alkyl alcohols) and therefore they are nonselective catalysts.

Due to the nonspecificity of the catalysts, a salt bridge must be employed to separate the anode and cathode compartments in order to increase the operating voltage of the electrochemical cell by separating anodic fuel from cross-reacting at the cathodic electrode. Theoretically, if selective catalysts were utilized at both electrodes, the polymer electrolyte membrane that acts as a salt bridge in a typical cell can be eliminated from the system. This can result in simplifying the electrochemical power system as well as the manufacturing procedure, which will result in lower production costs. Elimination of resistance that is associated with the polymer electrolyte membrane results in an increase in ion conductivity that results in higher power density outputs.

FUEL CELLS

In 1839, William Grove demonstrated the first fuel cell, which employed a very simple and basic system where generation of electricity was accomplished by supplying hydrogen and oxygen to two separate electrodes that were immersed in sulfuric acid [6]. A schematic of a hydrogen/oxygen fuel cell is shown in Figure 12.1. In a fuel cell, the respective fuel is oxidized at the anode producing and discharging electrons to an external circuit, which transfers them to the cathode where they are utilized along with discharged proton to reduce oxygen to water. Fuel cells (like batteries) consist of two electrodes and at least one electrolyte; but unlike batteries, their lifetime is much longer due to the fact they are easily recharged by the addition of more fuel to the anode chamber.

Batteries store energy, so they eventually expire, whereas a fuel cell is an energy conversion device that will produce power as long as fuel is supplied. Fuel cells were first developed for use in space vessels due to the demand for higher-power density and long-term power supply, which could not be delivered by traditional batteries [7]. Electrical power in outer space was required for operation of scientific data collection and transmission instrumentation to transmit information back to earth. The average battery has a lifetime up to 30 hours, which is relatively short for operations in outer space, where demand is at least 200–300 hours of constant supplied power.

The general interest in fuel cells is due to their potentially high efficiency. Efficiency is the ratio of energy produced to the amount of energy supplied, which is always higher than the energy produced. Conversion processes of one form of energy to another are never 100%. Therefore, the energy lost is actually converted to another form of energy, since the First Law of Thermodynamics declares that

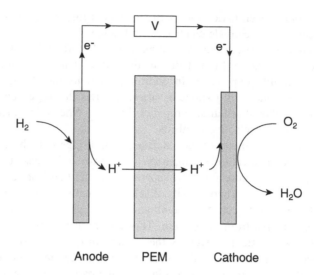

FIGURE 12.1 Schematic of a hydrogen/oxygen fuel cell.

energy is conserved so it is neither created nor destroyed [5]. Energy loss goes either into the form of sound, light, and/or, commonly, heat.

Applications of fuel cells are enormous and very diverse, but can be categorized by power output requirements. For example, high power is needed for industrial applications, medium power for domestic installations, and low power is necessary for certain kinds of vehicles and for use in space as well as portable power devices [7]. Different fuel cell types are distinguished by the electrolyte utilized and have varied applications due to differences in stability and strengths of power supply as well as lifetime and working conditions. There are six different types of fuel cells that all have different applications and operational circumstances; these are described in Table 12.1. They are alkaline fuel cell (AFC), proton exchange membrane fuel cell (PEMFC), direct methanol fuel cell (DMFC), phosphoric acid fuel cell (PAFC), molten carbonate fuel cell (MCFC), and solid-oxide fuel cell (SOFC) [8].

BIOFUEL CELLS

The rise in use of portable electronic devices has been increasing steadily in the United States and abroad over the past few years and most likely will continue to increase over the years to come as the population becomes more dependent on multifunctional portable electronics. Harvesting energy from renewable resources has become an important focus in order to eliminate our dependency on oil and other nonrenewable resources necessary as primary power sources. It is well known that industrialized nations are the highest energy consumers and that there is a correlation between energy consumption and status of economic and technological development [9]. About 65% of the world's primary energy was consumed in 1992 by industrialized countries and some of the more populated

TABLE 12.1
Comparison of the Six Traditional Types of Fuel Cells

Fuel Cell Type	Mobile Ion	Operating Temperature	Applications and Notes
Alkaline (AFC)	OH^-	50–200°C	Used in space vehicles, e.g., Apollo, shuttle
Proton exchange membrane (PEMFC)	H^+	30–100°C	vehicles and mobile applications, and for lower-power systems
Direct methanol (DMFC)	H^+	20–90°C	Suitable for portable electronics systems of low power, running or long times
Phosphoric acid (PAFC)	H^+	~220°C	Large numbers of 200-kW systems in use
Molten carbonate (MCFC)	CO_3^{-2}	~650°C	Suitable for medium- to large-scale systems, up to MW capacity
Solid oxide (SOFC)	O^{2-}	500–1000°C	Suitable for all sizes of CHP systems, 2-kW to multi-MW

countries, while developing areas consume more biomass energy such as wood and wood wastes.

The demand for energy is slowly increasing with developing technology, which in turn explains the extreme situation of the United States. The United States has only 5% of the world's population and yet consumes about one quarter of the total global primary energy. The sources for energy production in the United States are usually obtained from coal, natural gas, and oil where oil is most common [9]. In order to minimize our dependency on oil, researchers are attempting to harvest energy from renewable resources, such as alcohols, sugars, fats, and other biologically derived materials.

Biofuel cells are electrochemical devices in which energy derived from biochemical reactions is converted to electrical energy by means of the catalytic activity of microorganisms and/or their enzymes. Unlike metal catalysts, biocatalysts are derived from biomatter, which is a renewable resource. Recent biofuel cell research has explored using enzymes as biocatalysts due to their availability and specificity. Enzymes are functional proteins whose purpose is to catalyze specific biochemical reactions by lowering the activation energy of the reaction, without undergoing a permanent chemical change itself. Enzymes can be manipulated and produced by genetic engineering or harvested and extracted from living organisms. Both means of acquiring enzymes are more cost effective than mining precious metals used as traditional catalysts. Biofuel cell catalysts are more environmentally friendly compared to heavy metal batteries due to the fact they naturally biodegrade. Another advantage of enzyme employment in biofuel cells is the enzyme specificity that pushes the fuel cell technology one step further. Specificity of the enzyme's fuel utilization eliminates the need for employment of a salt bridge and therefore simplifies the fuel cell system [4].

The first biofuel cell was demonstrated by Potter in 1912 by employing glucose and yeast to obtain electrical energy [10]. This concept inspired scientists to investigate the metabolic pathways of power production [10]. Early biofuel cells employed microorganisms to oxidize the fuel for electricity generation; however, due to the slow mass transport of fuel across the cell wall, power densities are too low for practical applications. State-of-the-art microbial fuel cells developed by Lovley have shown greater than 40-day lifetimes, but power densities of 0.0074 mA/cm^2 [11].

More recently, enzyme-based fuel cells were constructed employing enzymes in the solution. These fuel cells had higher power densities due to the elimination of cell walls that slowed the mass transport; however, their lifetime only extended from hours to a few days because of the enzyme's stability. In contrast, higher power densities have been obtained with enzymatic fuel cells reaching up to 0.28 mW/cm^2 for a glucose/oxygen membraneless biofuel cell at room temperature [12] and 0.69 mW/cm^2 for a methanol/oxygen biofuel cell with a polymer electrolyte membrane [13]; however, enzymatic fuel cells are plagued with low lifetimes ranging from two hours [14] to seven days [15]. Table 12.2 depicts a brief history of biofuel cell technology.

Enzymes have been shown to be effective biocatalysts for biofuel cells compared to microbial biofuel cells. However, enzymes are very delicate catalysts. The optimal activity of enzymes depends on their three-dimensional configuration, which can be denatured with slight changes in pH or temperature. Therefore, it is necessary to develop an immobilization technique that will keep the enzyme active at the electrode surface in its optimal working conditions.

ENZYME IMMOBILIZATION TECHNIQUES

Over the last decade, there has been substantial research on immobilizing enzymes at electrode surfaces for use in biofuel cells [12,14–15]. These immobilization strategies have been successful at increasing biofuel cell lifetimes to 7–10 days

TABLE 12.2
Timeline of Improvements in Biofuel Cell Technology

	1960s	1980s	1990s	2003	2005
Stage of technology	Biofuel cells conceived living bacteria	Employed isolated enzymes in solution	Immobilized enzymes at electrode surface	Stabilize enzymes by casting in a polymer	Eliminating the mediator
Current density (mA/cm^2)	0.0002	0.52	0.83	5.00	9.28
Open circuit potential (V)	0.75	0.3	0.8	0.6	1.0
Lifetime	1–3 hours	1–3 days	3–14 days	45 days	>1 year

[15]. Therefore, one of the main obstacles that is still plaguing enzyme-based biofuel cells is the ability to immobilize the enzyme in a membrane at the electrode surface that will extend the lifetime of the enzyme and form a mechanically and chemically stable layer, while not forming a capacitive region at the electrode surface.

The problem associated with bioelectrodes as reported in the literature is ineffective techniques for enzyme immobilization [16]. The most common techniques used are sandwich [17] or wired [16,18]. However, sandwich and wired techniques still leave the enzyme exposed to the matrix, so the enzyme's three-dimensional configuration can change due to the harsh physical and chemical forces resulting in the loss of optimal enzymatic activity [14,16,18,19].

To solve these issues and offer a more stable enzyme immobilization, researchers have employed a micellar polymer (Nafion®). Nafion® is a perfluorinated ion exchange polymer that has excellent properties as an ion conductor and has been widely employed to modify electrodes for a variety of sensor and fuel cell applications. The molecular structure of Nafion® is shown in Figure 12.2. Nafion® is a cation exchange polymer that has superselectivity against anions. Nafion® also preconcentrates cations at the electrode surface and serves as a protective coating for the electrode. A simple approach to obtain selective electrodes is performed by solvent casting of the Nafion® polymer directly onto the electrode surface. Nafion® can be employed for enzyme immobilization in three different ways by employing either the wired technique, sandwich technique, or entrapment technique [20].

WIRED TECHNIQUE

The "wired" technique is most commonly used today for enzyme immobilization and can be employed with Nafion® polymers, as shown in Figure 12.3. However, this approach decreases the activity of the enzyme due to the change in the three-dimensional configuration of the enzyme that results from covalent bonding between the enzyme and the polymer. Another problem associated with this technique is that the enzyme is still subjected to the chemical environment of the

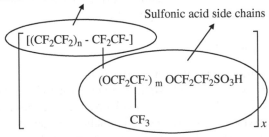

Teflon-based fluorocarbon backbone

Sulfonic acid side chains

$$\left[\; [(CF_2CF_2)_n - CF_2CF-] \atop (OCF_2CF-)_m \; OCF_2CF_2SO_3H \atop CF_3 \right]_x$$

Nafion® 117 m = 1 n = 6, 7...14

FIGURE 12.2 Structure of Nafion polymer.

Nafion Coating

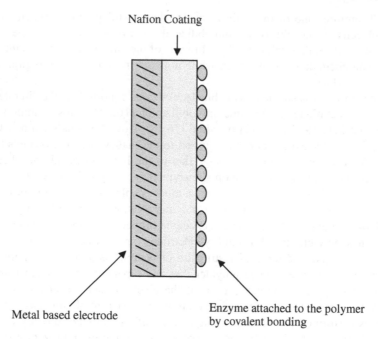

Metal based electrode

Enzyme attached to the polymer
by covalent bonding

FIGURE 12.3 Enzyme immobilization by "wiring" technique.

matrix and not protected from its surroundings. Therefore, the enzyme can be easily denatured, and this limits the lifetime of the enzymatic catalytic activity.

SANDWICH TECHNIQUE

A second type of enzyme immobilization employing Nafion® is the sandwich technique in which the enzyme is trapped in between the polymer and the electrode surface, as shown in Figure 12.4. This is accomplished by simply casting the enzyme solution onto an electrode surface before casting the Nafion® suspension. Sandwich techniques are powerful and successful for enzyme immobilization; however, the enzyme's optimal activity is not retained due to the physical distress applied by the polymer. In addition to this, the diffusion of analyte through the polymer is slowed limiting its applications.

ENTRAPMENT TECHNIQUE

The third technique for enzyme immobilization is employing micellar polymer Nafion® for enzyme entrapment within the pore structure of the membrane, as shown in Figure 12.5. However, commercial Nafion® has not been successful at immobilizing enzymes at the surface of biofuel cell electrodes because Nafion® forms an acidic membrane that decreases the lifetime and activity of the enzyme. Researchers have been successful in maintaining the activity of glucose oxidase enzymes immobilized in Nafion® by diluting the Nafion® suspension [20];

Enzyme casting layer

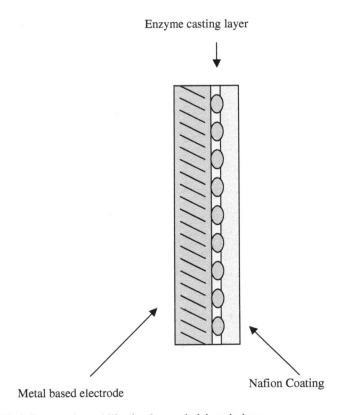

Metal based electrode

Nafion Coating

FIGURE 12.4 Enzyme immobilization by sandwich technique.

however, this approach did not form stable and uniform films. The most recent method employed by Karyakina and coworkers was to neutralize the Nafion® casting solution and dilute the solution to a lesser degree in ethanol; however, both of these approaches have problems with maintaining activity of enzymes for extended times. As the pH environment in the solution around the Nafion® membrane decreases, protons will exchange back into the membrane and re-acidify the membrane [20].

NAFION® MODIFICATION

The technique developed by Minteer et al. [20–22] involves modifying Nafion® with quaternary ammonium bromide salts. This technique provides an ideal environment for enzyme immobilization due to the biocompatibility and structure of the micellar pores. This method helps to keep the enzyme at the electrode surface as well as maintain high enzymatic activity and protect the enzyme from the surrounding environment. Previous studies by Schrenk et al. have shown that mixture-cast films of quaternary ammonium bromide salts and Nafion® have increased the mass transport of small analytes through the films and decreased the selectivity of the membrane against anions [21].

Enzyme entrapped within a micellar
polymer

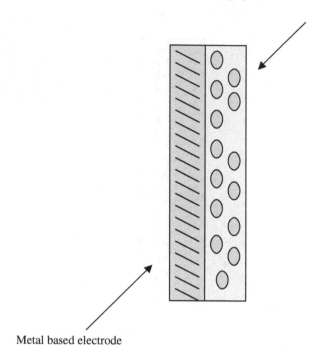

Metal based electrode

FIGURE 12.5 Enzymes immobilized by entrapment.

Quaternary ammonium bromide salts have a higher affinity to the sulfonic acid side chain than the proton; therefore, they can be utilized to modify the polymer and extend the enzymes lifetime because protons are less likely to exchange back into the membrane and reacidify it. A much higher preference to the quaternary ammonium bromide salts than to the proton has been shown by titrating the number of available exchange sites to protons in the membranes [21]. Due to the fact that quaternary ammounium bromide salts are larger in size than protons, the micellar structure will also be enlarged to facilitate enzyme entrapment.

Immobilizing enzymes in micellar pores will eliminate the issue of covalent bonding, which "wired" techniques are plagued with, because the process can buffer the pH of the membrane for optimal enzyme catalytic activity. In addition, the pore structure also provides a protective and restrictive 3D pore, unlike "wired" techniques where an enzyme is freely subjected to the surroundings and can be easily denatured if introduced to a harsh environment. Also, quaternary ammonium bromide salts have similar hydrophobicity as the enzymes. Nafion® modified with quaternary ammonium bromide salts will not only provide a buff-ered micellar structure for easier enzyme immobilization, but will also retain the electrical properties of unmodified Nafion® as well as increase the mass transport

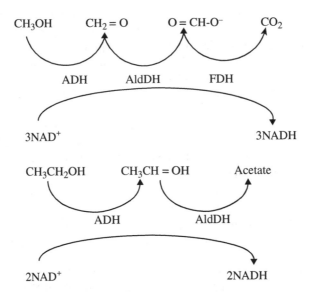

FIGURE 12.6 Oxidation schemes for methanol (top) and ethanol (bottom).

flux of ions and neutral species through the membrane minimizing problems associated with slow diffusion [20].

NAD⁺-DEPENDENT ALCOHOL DEHYDROGENASE BIOFUEL CELLS

The bioanode of the biofuel cell is the electrode at which the fuel is utilized by enzymes to produce electrons and protons, which are then utilized by enzymes of biocathodes to reduce O_2 to H_2O. Alcohol-based enzymatic systems that have been chosen most frequently for the bioanode involve NAD⁺-dependent alcohol dehydrogenase (ADH), which oxidizes alcohols to aldehydes. This enzyme can be employed with aldehyde dehydrogenase to further oxidize the aldehyde. Literature reports of alcohol biofuel cells are limited to only two alcohol-based enzymatic schemes. They are for methanol and ethanol and are shown in Figure 12.6. Methanol is oxidized to formaldehyde by alcohol dehydrogenase and then the formaldehyde is oxidized to formate by formaldehyde dehydrogenase. The formate is completely oxidized to carbon dioxide by formate dehydrogenase. The ethanol system involves oxidizing ethanol to acetaldehyde by alcohol dehydrogenase and then oxidizing the acetaldehyde to acetate by aldehyde dehydrogenase.

All of these enzymatic systems require NAD⁺ as a coenzyme/cofactor and the reduced form (NADH) is the hydrogen source at the electrode surface. However, NADH has a high overpotential at most typical electrode surfaces (platinum, carbon, etc.), so an electrocatalyst layer is necessary to decrease the potential and increase the power output. The problem with electrocatalyst layers is that they are not as conductive as carbon or most metals and they add an extra complexity to the system that makes forming high surface area bioanodes difficult. A variety of electrocatalysts have been used, but at this stage, there is no optimal electro-catalyst. Palmore and coworkers have employed a diaphorase/benzyl viologen system that has shown good thermodynamics properties, but poor lifetimes [13]. Minteer and coworkers have employed methylene green as the electrocatalyst layer due to its optimal electrocatatlytic properties. Poly(methylene green) pre-pared via electropolymerization has been shown to be an electrocatalyst for NADH [23].

The second problem with an NAD$^+$-dependent bioanode is the instability of the NAD$^+$/NADH couple in the membrane. When ethanol is oxidized to acetate, the NAD$^+$ is converted to NADH. It is simple to electrostatically immobilize NAD$^+$ in the bioanodes membrane, but NAD$^+$ has a short lifetime in solution and a limited lifetime in the membrane. NAD$^+$ is only stable in solution for a few hours, but it can be stabilized for up to 45 days in the membrane. Dehydrogenase enzymes are stable for much longer (>6 months) in the membrane, so it is necessary to employ a coenzyme that is stable for at least as long as the enzyme is stable.

Akers et al. have tested ethanol-based biofuel cells fabricated using bioanodes containing NAD$^+$-dependent ADH immobilized in a modified Nafion® membrane as discussed above and cathodes formed from ELAT electrodes with 20% Pt on Vulcan XC-72 (E-Tek). These bioanodes can function for greater than 30 days [23]. The test cell contains an anode solution of 1.0 mM ethanol in pH 7.15 phosphate buffer and a cathode solution containing pH 7.15 phosphate buffer saturated with dissolved oxygen. The two solutions are separated by a Nafion® 117 membrane. The ethanol-based biofuel cells have had open-circuit potentials ranging from 0.61 to 0.82 V at 20°C and have maximum power densities of 1.12 mW/cm^2 [23]. This is a 16-fold increase in power density versus the state-of-the-art biofuel cell developed by Heller and co-workers [24]. The milestone that was required to further develop a biofuel cell is to eliminate the need for the electro-catalyst layer, poly(methylene green). This will be done by replacing NAD$^+$-dependent ADH with PQQ-dependent ADH.

PQQ-DEPENDENT ALCOHOL DEHYDROGENASE BIOFUEL CELLS

More recent research in developing long-term stability in biofuel cell systems has focused on studying a new enzymatic system. Initial studies have been successful in utilizing PQQ-dependent ADH as a catalyst at the anode of a biofuel cell. Pyrroloquinoline quinone (PQQ)-dependent alcohol dehydrogenase (ADH) has been chosen to replace NAD$^+$-dependent ADH in order to extend the lifetime and simplicity of the fuel cell. PQQ is the coenzyme of PQQ-dependent ADH, and it remains electrostatically attached to PQQ-dependent ADH; therefore, the enzyme and coenzyme will remain in the membrane leading to an increased lifetime and activity for the biofuel cell. Also, PQQ-dependent ADH possesses desirable electrochemistry (it has the ability to transition between its oxidized and reduced state). The coenzyme PQQ has quasireversible electrochemistry and a low overpotential at an unmodified carbon electrode. This eliminates the need for an electrocatalyst layer, thereby simplifying the process of forming high-surface-area bioanodes.

PQQ-dependent ADH bioanodes fabricated and tested by Treu et al. using the same procedures and methods as for the NAD^+-dependent ADH system had desirable results. The PQQ-dependent ADH enzymatic system has shown lifetimes of greater than 1 year of continuous use, open circuit potentials of 1.0 V and power densities of up to 3.61 mW/cm^2 for a 1.0 mM ethanol solution at room temperature in a static system [25].

The PQQ-dependent ADH-based biofuel cell outperformed the NAD^+-dependent ADH-based fuel cell in all aspects of traditional systems. For the traditional system of PQQ-dependent ADH anode coupled with a traditional platinum cathode, the PQQ-based enzymatic system increased the overall lifetime by greater than 711%, increased open circuit potential by 67%, and increased power density by 251% [25]. There was no correlation between power and different pHs for the PQQ-dependent ADH biofuel cell [26].

PQQ-dependent ADH has optimum selectivity for ethanol, but will oxidize other alcohols. Similar fuel cells have been developed for methanol, butanol, and propanol using PQQ-dependent ADH bioanodes. The performance of those fuel cells is shown in Table 12.3. PQQ-dependent ADH is a desirable substitute for a biocatalyst at the anode of a biofuel cell. Eliminating the poly(methylene green) layer simplifies the fabrication of a bioanode and lowers the IR drop leading to an increase in performance. Research has shown that PQQ-dependent ADH enables simple and timely electrode fabrication, along with impressive open circuit potentials, current densities, power densities, and lifetime for a complete ethanol/oxygen biofuel cell.

MEMBRANELESS BIOFUEL CELLS

Since enzymes are highly selective, there are limited problems associated with fuel crossover from the anode to the cathode in a biofuel cell. If an anode and cathode are both selective, then a polymer electrolyte membrane is no longer required to separate the anode and cathode solutions. Topcagic et al. have developed the first ethanol membraneless biofuel cell. At the cathode, bilirubin oxidase has been chosen to replace platinum as the reducing catalyst to increase specificity of the cathode. A schematic showing the simplicity of a membraneless biofuel cell can be seen in Figure 12.7.

Alternatives for platinum found in the literature typically use laccase enzyme as studied by Heller's Group [27]. Laccase lowers the power of a biofuel cell due to the maximum turnover rate of laccase occuring at pH 5.0 and deactivation in the presence of chlorine ions. Bilirubin oxidase has been chosen as a catalyst for future studies, because it has optimum performance in a physiological environment (near-neutral pH and presence of various ions). The second problem associated with many biocathodes in the literature is that electrodes are osmium-based creating a toxicity hazard to the surrounding environment. Topcagic et al. have replaced the osmium-based mediator with a ruthenium-based complex of

TABLE 12.3
Performance Data for a Variety of Biofuel Cell Configurations (ADH is alcohol dehydrogenase and AldDH is aldehyde dehydrogenase)

Biofuel Cell	Fuel	Open Circuit Potential (V)	Maximum Current Density (mA/cm^2)	Maximum Power Density (mW/cm^2)	Lifetime (days)
NAD$^+$-dependent ADH anode with platinum cathode	Ethanol	0.60	—	1.16	45
NAD$^+$-dependent ADH and AldDH anode with platinum cathode	Ethanol	0.82	—	2.04	45
NAD$^+$-dependent ADH anode with biocathode	Ethanol	0.82	2.23	0.95	20
Membraneless NAD$^+$-dependent ADH anode with biocathode	Ethanol	0.95	6.10	2.67	30–60
PQQ-dependent ADH anode with platinum cathode	Methanol	0.79	3.37	1.98	—
	Ethanol	1.00	8.79	3.62	>365
	Propanol	0.51	2.51	1.63	—
	Butanol	0.55	1.80	1.05	—
Membraneless PQQ-dependent ADH anode with biocathode	Ethanol	1.04	8.47	2.44	>157

similar structure that is less toxic and has a higher self-exchange rate. A third problem associated with anodes and cathodes in the literature is a technique of immobilizing the enzyme at the electrode surface. Literature enzyme immobilization employs covalent bonding of the enzyme to the surface of the electrode or to the mediator. This method does not protect the enzyme from its surroundings and its optimal activity is lowered due to the conformational change that resulted from physically attaching the enzyme to the surface to the electrode [23]. However, instead of physically attaching the enzyme, Topcagic et al. have immobilized it in modified Nafion®; therefore optimum enzyme activity is retained and the enzyme is protected from the surrounding environment.

The membraneless biofuel cell operates at room temperature, which varies from 20–25°C in a phosphate buffer pH 7.15 containing 1.0 mM ethanol. Since the polymer electrolyte membrane has been eliminated, the electrodes have to be specific enough to work in same compartment. Both electrodes, bioanode and biocathode, consisted of immobilized enzyme casting solution at the surface of the 1-cm^2 carbon fiber paper. The performance of NAD$^+$-dependent ADH and PQQ-dependent ADH bioanodes coupled to biocathodes was also studied and is

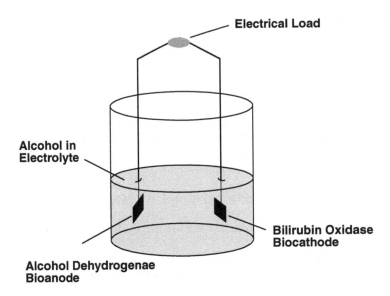

FIGURE 12.7 Membraneless biofuel cell schematic.

summarized in Table 12.3. The bioanode used for most studies had PQQ-dependent alcohol dehydrogenase enzymes, while the biocathode had bilirubin oxidase, bilirubin with $Ru(bpy)_3{}^{+2}$ immobilized at the surface of the electrode. Maximum open circuit potential is 1.04 V with maximum current density of 8.47 mA/cm^2 [28]. For the membraneless system comprised of a PQQ-dependent ADH anode and bilirubin oxidase biocathode, the fuel cell has an increased lifetime of greater than 353%, increased open circuit potential of 15%, and increased power density of 97% compared to the NAD$^+$-dependent ADH bioanode [28].

CONCLUSIONS

Research has succeeded in increasing the stability of enzymes at the electrode surface, which in turn increases the open circuit potential, current and power of a biofuel cell. Also, by eliminating the need for an electrocatalyst layer and a polyelectrolyte membrane, it reduces the cost of production of the current biofuel cell and simplifies fabrication. Replacing NAD$^+$-dependent ADH with PQQ-dependent ADH is a step toward reaching the goal of increasing the overall lifetime of the biofuel cell for future use in multiple power applications. The most important phenomenon to examine is the increase in lifetime of the membraneless system with PQQ-dependent bioanodes. The PQQ-dependent bioanodes are more stable than NAD$^+$-dependent bioanodes. In addition to better data results, PQQ-dependent ADH serves as an invaluable replacement for NAD$^+$-dependent ADH due to the simplicity it offers in bioanode fabrication.

REFERENCES

1. Gorisch, H., The Ethanol Oxidation System and its Regulation in Pseudomonas Aeruginosa, *Biochimica et Biophysica Acta*, 98, 1647, 2003.
2. Ikeda, T., Kato, K., Tatsumi, H. and Kano, K., Mediated Catalytic Current for the Oxidation of Ethanol Produced by Acetobacter Aceti Cells Suspended in Solution, *J. Electroanal. Chem.*, 265, 440, 1997.
3. Molinari, F., Villa, R., Manzoni, M. and Aragozzini, F., Aldehyde Production by Alcohol Oxidation with Gluconobacter Oxydans, *Applied Microbiology and Biotechnology*, 43, 989, 1995.
4. Moore, C.M., *From Macro to Micro Biofuel Cells*, thesis, Saint Louis University, 2004.
5. Brown, T.L., LeMay, H.E., Jr. and Bursten, B.E., *Chemistry*, Pearson Education, Inc., New Jersey, 2003.
6. Williams, K.R., *An Introduction to Fuel Cells*, Elsevier Publishing Company, Amsterdam, 1966.
7. McDougall, A., *Fuel Cells*, John Wiley and Sons, New York, 1976.
8. Dicks, A., *Fuel Cell Systems Explained*, John Wiley and Sons, Ltd., United Kingdom, 2003.
9. Klass, D.L., *Biomass fo Renewable Energy, Fuels, and Chemicals*, Academic Press, San Diego, 1998.
10. Williams, K.R., *An Introduction to Fuel Cells*, Elsevier Publishing Company, New York, 1966.
11. Chaudhuri, S. and Lovley, D.R., Electricity generation by direct oxidation of glucose in mediatorless microbial fuel cells, *Nature Biotechnology*, 21, 1229, 2002.
12. Mano, N., Mao, F. and Heller, A., Characteristics of a miniature compartmentless glucose-O_2 biofuel cell and its operation in a living plant, *J. Am. Chem. Soc.*, 125, 6588, 2003.
13. Palmore, G.T., Bertschy, H., Bergens S.H. and Whitesides, G.M., A Methanol/Dioxygen Biofuel Cell That Uses NAD+-Dependent Dehydrogenase as Catalysts: Application of an Electro-Enzymatic Method to Regenerate Nicotinamide Adenine Dinucleotide at Low Overpotentials, *Journal of Electroanalytical Chem.*, 443, 155, 1998.
14. Mano, N., Kim, H.H. and Heller, A. On the Relationship between the Charactericstics of Bilirubin Oxidase and O2 Cathodes Based on Their "Wiring," *J. Phys. Chem. B*, 106, 8842, 2002.
15. Mano, N., Mao, F., Shin, W., Chen. T. and Heller, A., A miniature biofuel cell operating at 0.78V, *Chem. Commun.*, 4, 518, 2003.
16. Barton, S.C., Kim, H., Binyamin, G., Zhang, Y. and Heller, A., Electroreduction of O_2 to Water on the "Wired" Laccase Cathode, *J. Phys. Chem. B*, 105, 11917, 2001.
17. Kim, E., Kim K., Yang, H., Kim, Y.T. and Kwak, J., Enzyme-Amplified Electrochemical Detection of DNA Using Electrocatalysis of Ferrocenyl-Tethered Dendrimer, *Anal. Chem.*, 75, 5665, 2003.
18. Mano, N., Mao, F. and Heller, A., A Miniature Biofuel Cell Operating in a Physiological Buffer, *J. Am. Chem. Soc.*, 124, 12962, 2002.
19. Kim, H., Zhang, Y. and Heller, A., Bilirubin Oxidase Label for an Enzyme-Linked Affinity Assay with O_2 as Substrate in a Neutral pH NaCl Solution, *Anal. Chem.*, 76, 2411, 2004.

20. Moore, C.M., Akers N.L., Hill, A.A., Johnson, Z.C. and Minteer, S.D., Improving the Envoroment for Immobilized Dehydrogenase Enzymes by Modifying Nafion with Tetraalkylammonium Bromides, *Biomacromolecules*, 5, 1241, 2004.

21. Schrenk, M.J., Villigram, R.E., Torrence, N.J., Brancato, S.J. and Minteer, S.D., Effects of mixture-casting Nafion with quaternary ammonium bromide salts on the ion-exchange capacity and mass transport in the membranes, *J. Membrane Science*, 203, 3, 2002.

22. Thomas, T.J., Ponnusamy, K.E., Chang, N.M., Galmore, K. and Minteer, S.D., Effects of annealing on mixture-cast membranes of Nafion and quaternary ammonium bromide salts, *J. Membrane Science*, 213, 55, 2003.

23. Akers, N.L., Moore, C.M. and Minteer, S.D., Development of Alcohol/Oxygen Biofuel Cells Using Salt-Extracted Tertabutylammonim Bromide/Nafion Membranes to Immobilize Dehydrogenase Enzymes, *Electrochimica Acta*, 50, 2521, 2005.

24. Chen, T., Barton, S.C., Binyamin, G., Gao, G., Zhang, Y., Kim, H.H. and Heller, A., A Miniature Biofuel Cell, *Journal of the American Chemical Society*, 123, 8630, 2001.

25. Treu, B.L. and Minteer, S.D., Improving the Lifetime, Simplicity, and Power of and Ethanol Biofuel Cell by Employing Ammonium Treated Nafion Membranes to Immobilize PQQ-Dependent Alcohol Dehydrogenase, *Polymeric Materials: Science and Engineering*, 92, 192, 2005.

26. Treu, B.L., The Development of a PQQ-Dependent Alcohol Dehydrogenase Bioanode, thesis, Saint Louis University, 2005.

27. Heller, A., Miniature Biofuel Cells, *Physical Chemistry Chemical Physics*, 6, 209, 2004.

28. Topcagic, S., Treu, B.L. and Minteer, S.D., Characterization of Oxygen Biocathodes Employing Tetrabutylammonium Bromide Treated Nafion Immobilization Membranes, *Polymeric Materials: Science and Engineering*, 92, 201, 2005.

13 Ethanol Reformation to Hydrogen

Pilar Ramírez de la Piscina and Narcís Homs
Inorganic Chemistry Department,
Universitat de Barcelona, Spain

CONTENTS

BACKGROUND

Energy is one of the main factors that must be taken into account when sustainable development of our society is envisioned because there is an intimate connection between energy, the environment and development. In response to the need for cleaner and more efficient energy technology, a number of alternatives to the current energy network have emerged. In this context, the general use of fuel cells for automotive purposes or stationary power generation is envisioned in the medium term. This is a promising advance in the production of electrical energy from chemical energy, since the efficiency of a fuel cell is much higher than that of a combustion engine.

The fuel most widely studied for use in a fuel cell is hydrogen. Although the ideal situation would be the production of hydrogen from water, using renewable energy sources (e.g., solar energy), this is unlikely to become extensively operative in the short to medium term. At present, hydrogen is mainly produced by steam reforming of fossil fuel-derived feedstock, mostly natural gas and naphtha.

The main objective of the steam reforming process is to extract the hydrogen from the substrate. From hydrocarbons, hydrogen is obtained via the general equation:

$$C_nH_{2n+2} + nH_2O \leftrightarrow nCO + (2n + 1)H_2 \quad \Delta H^\circ > 0$$

Then, the production of hydrogen is completed by the successive water gas shift reaction (WGSR):

$$CO + H_2O \leftrightarrow CO_2 + H_2 \quad \Delta H^\circ = -41.1 \text{ kJ mol}^{-1}$$

Both reactions can only be carried out in a practical way by catalytic means. The steam reforming reaction is endothermic and the real amount of energy required depends on both the stability of the substrate to be reformed and the ability of the catalyst to activate and transform the substrate into the products. The WGSR is slightly exothermic, and the forward reaction is not favored at the temperature used for steam reforming, which is higher than 1000 K for CH_4. Therefore, the overall process requires the use of different catalysts, which operate under different reaction conditions in separate reactors. In the case of natural gas and naphtha, many years of industrial practice have led the total process to become technologically mature. However, if a strong increase in the demand for hydrogen is contemplated, some advanced research and development in catalysis and technology would still be needed in the next few years [1,2].

On the other hand, society has become environmentally conscious and sensitive to its oil dependency because petroleum is likely to become scarce and expensive and the reserves are concentrated in a few countries. If a long-term global solution is envisioned, other, nonfossil-derived fuels, which are renewable and environmentally friendly must be contemplated for the supply of hydrogen. In this context, ethanol is a very promising alternative. As has been stated in previous chapters, ethanol, which can be considered a renewable and ecofriendly hydrogen carrier, can be produced from a large variety of biomass-based sources.

The catalytic steam reforming of ethanol may provide up to 6 moles of hydrogen per mol of ethanol reacted:

$$CH_3CH_2OH + 3H_2O \leftrightarrow 2CO_2 + 6H_2 \quad \Delta H^\circ = 173.4 \text{ kJ mol}^{-1}$$

If the primary production of CO is considered in the steam reforming of ethanol, the WGS reaction must be taken into account. The overall process, then, will be the combination of both reactions:

$$CH_3CH_2OH + H_2O \leftrightarrow 2CO + 4H_2$$

$$CO + H_2O \leftrightarrow CO_2 + H_2$$

TABLE 13.1

Several Thermodynamic Constants of Ethanol Steam Reforming [4]; $CH_3CH_2OH + 3H_2O \leftrightarrow 2CO_2 + 6H_2$

T(K)	ΔH (kJ/mol ethanol reacted)	ΔH (kJ/molH_2 generated)	K_p
298.15	173.36	28.89	$5.49\ 10^{-13}$
600	193.95	32.33	$5.33\ 10^4$
1000	208.80	34.80	$5.32\ 10^{11}$

Although globally the reaction releases 2 moles of carbon dioxide, the total process is almost neutral from the point of view of CO_2 generation, since it may be assumed that the CO_2 produced is consumed in biomass growth. Consequently, the use of the steam reforming of ethanol as a source of hydrogen can contribute to the global reduction of CO_2 emissions. Moreover, other emissions of greenhouse or polluting gases such as hydrocarbons and NOx could also be mitigated.

As we have just said, the reaction of ethanol steam reforming is highly endothermic. However, theoretical and experimental studies have shown that ethanol steam reforming can take place at temperatures above 500 K [3]. Table 13.1 shows that relatively high values of equilibrium constant (Kp) can be achieved for temperatures of over 600 K. On the other hand, it is worth mentioning that, in this case, the energy required per mol of hydrogen generated (Table 13.1) is lower than half of that required to obtain hydrogen from the steam reforming of hydrocarbons. As an example, values of H (kJ per mol of hydrogen generated) at 600 K can be considered; 32.33 kJ must be supplied when H_2 is obtained from ethanol, and 72.82 kJ if methane is used [4].

An issue of major importance in ethanol steam reforming is the development of catalysts that operate with high levels of activity, selectivity, and stability. Several products that can be formed under reaction conditions could need other experimental conditions to be reformed. Consequently, the total process leading to an effluent that mainly contains H_2 and CO_2 and is free of undesirable products may be complex. Depending on the reaction conditions and catalyst used, the following reactions could contribute to a low selectivity of the process, among others:

$$CH_3CH_2OH \rightarrow CH_3CHO + H_2$$

$$CH_3CH_2OH \rightarrow CH_2CH_2 + H_2O$$

$$CH_3CH_2OH \rightarrow CH_4 + CO + H_2$$

$$CH_3CHO \rightarrow CH_4 + CO$$

$$CO_x + (2 + x)H_2 \rightarrow CH_4 + xH_2O$$

Thus, after the steam reforming, an additional purification of the effluent could be necessary, but this will depend on the fuel cell to be fed. For hydrogen operating in a polymer membrane fuel cell (PEMFC) or phosphoric acid fuel cell (PAFC) the limit of CO concentration in the fuel is 50 ppm and 0.05%, respectively [5]. These low CO concentrations may be achieved by subsequent catalytic selective oxidation or methanation processes or by the use of H_2 selective membranes. An additional purification of the reformed effluent might be unnecessary when a molten carbonate fuel cell (MCFC) or a solid-oxide fuel cell (SOFC) is used. Both fuel cells, which operate at high temperatures, may convert impurities of CH_4 and CO in the anode chamber [5,6].

Moreover, to make the steam reforming of ethanol operative in practice it must be energetically integrated with other exothermic processes, e.g., combustion or partial oxidation, which may supply the energy required for the steam reforming.

In the following sections, some propositions for globally energetically integrated processes and the main catalytic systems used to date for the different reactions will be analyzed. Finally, relevant perspectives of the development of the ethanol reformation to hydrogen in the near future will be presented.

ENERGETICALLY INTEGRATED ETHANOL REFORMING PROCESSES

This section will show several options of integration of the steam reforming of ethanol in overall processes that are energetically favored. Some of these processes have been proposed for the production of hydrogen from biomass-derived ethanol.

Figure 13.1 shows a schematic based on that proposed by Verikyos and colleagues [7], in which bioethanol could come from different sources: plants, agroindustrial residues, and the organic fraction of municipal solid waste. Besides bioethanol, biogas is produced. Aqueous solutions of ethanol initially generated could be concentrated prior to the steam reforming by a conventional distillation procedure. The efficiency of the overall process has been estimated to be twice that of the process producing electricity by a conventional method from biomass combustion [7].

This is a heat-integrated process, in which the heat necessary in the steps of distillation and steam reforming reactions is supplied by the heat evolved from several exothermic chemical reactions. Among these, the electrochemical reaction in the fuel cell may contribute to the energy balance of the total process. For instance, in the case of an MCFC, one of the limiting factors of the technology for a high-yield operation is the recovery of heat generated at the anode [8]:

$$CO_3^{2-} + H_2 \rightarrow CO_2 + H_2O + 2e^- \quad \Delta H < 0.$$

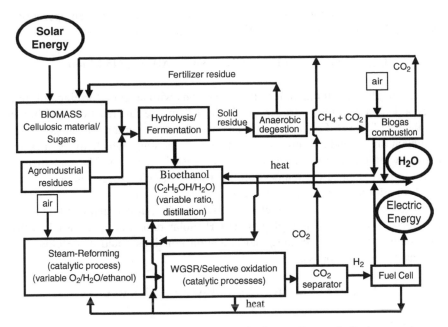

FIGURE 13.1 Renewable ethanol reformation to hydrogen. Energetically integrated process using a fuel-cell.

The use of this heat to produce H_2 by steam reforming may optimize the operation of the cell, and the direct contact of catalyst with vapor carbonates may be avoided by placing a chamber with the catalyst adjacent to the anode.

In MCFC and SOFC fuel cells, the catalyst may be placed in the anode compartment and then the internal steam reforming of ethanol occurs. In this case, formation of carbon residues must be avoided since the electrode structure may break down [9]. In this respect, it must be highlighted that the presence of some of the by-products, such as ethylene, has been related to the deactivation of catalysts by carbon deposition.

Moreover, for stationary applications, the use of heat generated by the combustion of methane (biogas) is proposed:

$$CH_4 + 2O_2 \rightarrow CO_2 + 2H_2O \quad \Delta H^\circ = -803 \text{ kJ mol}^{-1}$$

Recently, catalytic methane combustion coupled to ethanol steam reforming across an autothermal wall, which eliminated heat transfer boundary layers, was reported [10]. On the other hand, if oxygen is introduced in the ethanol:water mixture, the coupling of two oxidation reactions may take place:

$$CH_3CH_2OH + 3O_2 \rightarrow 2CO_2 + 3H_2O \quad \Delta H^\circ = -1280 \text{ kJ mol}^{-1}$$

$$H_2 + 1/2O_2 \rightarrow H_2O \quad \Delta H^\circ = -242 \text{ kJ mol}^{-1}$$

By tuning the amount of oxygen, using air as a carrier and an appropriate catalyst, the catalytic partial oxidation of ethanol can be effective:

$$CH_3CH_2OH + 3/2O_2 \rightarrow 2CO_2 + 3H_2 \quad \Delta H° = -545 \text{ kJ mol}^{-1}$$

Although the combination of these reactions with steam reforming will produce a decrease in the hydrogen yield, all of them are fast reactions and highly exothermic [11,12]. For mobile applications, there are certain advantages to the combination of endothermic and exothermic reactions, which overcomes the difficulty of quenching the exothermic process by cooling. The endothermic reaction may control the temperature of the exothermic process, alleviating the need for additional cooling equipment.

Thus, the operation of the global process under autothermal conditions can be proposed [13,14]. The $O_2/H_2O/C_2H_5OH$ molar ratio can be adjusted so that the combined steam reforming and oxidation reactions could come close to a thermally neutral process:

$$CH_3CH_2OH + xO_2 + (3 - 2x)H_2O \rightarrow 2CO_2 + (6 - 2x)H_2$$

for which

$$\Delta H° = ((3 - 2x)/3)173 - (x/1.5)545) \text{ kJ mol}^{-1}$$

Thermodynamic calculations under autothermal conditions, considering CH_3CH_2OH, O_2, and H_2O as reactants and H_2, CO, and CO_2 as products, have recently been carried out [4]. Other thermodynamically possible reaction products, such as methane, ethylene, etc. were not taken into account. Results for different fuels (methane, methanol, dimethylether, and gasoline) have been compared. The calculations lead to a maximum hydrogen content in the product of 41–43% when ethanol is used as fuel. This maximum is achieved between 530–600 K with a water/ethanol molar ratio of 1.6–2.9; under these conditions, the CO content in the effluent is approximately 5–10% [4].

Several studies are centered on the optimization of fuel cells that are fueled by H_2 produced in the ethanol steam reforming [3,15–17]. Ethanol has been claimed to be a promising alternative to CH_4 as a source of H_2 for these systems, and an efficiency for the SOFC system of 94% has been calculated for ethanol, which compares well with the efficiency values calculated for CH_4, 96%; methanol, 91%; and gasoline, 83% [16].

CATALYTIC SYSTEMS

Steam reforming reactions are catalyzed by metals of Groups 8–10 of the Periodic Table, nickel being preferred for industrial applications [5]. However, early studies on ethanol steam reforming were carried out over copper-based catalysts [18].

They were previously used extensively for methanol steam reforming because they were well-known catalysts for methanol synthesis and were available and highly cost effective. Moreover, copper is a very appropriate catalyst for dehydrogenation and WGS reactions, and both reactions may be involved in the total process of ethanol steam reforming. However, over copper-based catalysts under steam reforming conditions, ethanol can be transformed to ethyl acetate or acetic acid. The former transformation takes place via a nucleophilic addition of ethoxide or ethanol species to acetaldehyde, while the formation of acetic acid takes place via the nucleophilic addition of OH⁻ or H_2O to acetaldehyde [18]. Both transformations are favored only at temperatures below 600 K, and their rate is lower than the dehydrogenation reaction [18,19]. Iwasa and Takezawa have already pointed out differences between the behavior of copper-based catalysts in methanol and ethanol steam reforming [18]. In the former, the dehydrogenation of methanol to formaldehyde was found to be the rate-determining step, and then carbon dioxide and hydrogen were selectively produced. For ethanol steam reforming, acetaldehyde was formed by ethanol dehydrogenation, but then ethyl acetate and acetic acid were produced. Consequently, copper shows a lower selectivity in ethanol steam reforming because of its effectiveness in the cleavage of C–C bonds. Other metals such as cobalt, nickel, or noble metals, which show greater ability to break C–C bonds [20], were preferred for steam reforming of higher alcohols than methanol. Thus, ethanol steam reforming has been mainly studied over supported metal catalysts of Groups 9 and 10. Several studies indicate the influence of the nature of the support on the behavior of the catalysts. The support itself can promote the formation of different products under steam reforming conditions, depending on its acidic/basic properties and redox characteristics [21]. Over acidic supports dehydration of ethanol to ethylene may occur. Once it is formed, ethylene would adsorb very strongly on the metal component of the catalyst and become a precursor of coke formation. In this case the neutralization of the acid centers of the support by introduction of alkaline additives prevents the formation of ethylene and consequently diminishes the catalyst deactivation. On the other hand, basic centers may be involved in the ketonization reaction:

$$2CH_3CH_2OH \rightarrow CH_3COCH_3 + CO + 3H_2$$

which occurs through several successive reactions such as dehydrogenation and aldol condensation. Consequently, a decrease of hydrogen yield and the formation of undesired by-product could occur.

In what follows, relevant aspects of nickel, cobalt, and noble metal catalytic systems, including catalysts for autothermal operation, will be discussed. However, before entering into a discussion of the behavior of specific catalysts in the process, general trends of mechanistic aspects will be introduced. Then, the behavior of different catalysts in the steam reforming of ethanol will be analyzed in the light of their behavior in the consecutive steps that must be completed in order to ensure the total process.

MECHANISTIC ASPECTS

As stated above, the overall reaction network of ethanol steam reforming is highly complex because numerous reactions may coexist in equilibrium in the experimental conditions in which the process is carried out. Moreover, both the active metallic phase and the support can interact with ethanol, and consequently the proposed pathway depends not only on the reaction conditions but also on the nature of the catalyst.

For Ni- [7,22], Pd- [23], Co- [24–26], and Rh-based [8,27,28] catalysts, it has been proposed that the first step is the dehydrogenation of ethanol:

$$CH_3CH_2OH \leftrightarrow CH_3CHO + H_2$$

and in several cases the reaction has been determined to occur via a surface ethoxide species (CH_3CH_2O-). Over Ni-, Pd-, and Rh-based catalysts the subsequent decomposition of acetaldehyde into CH_4 and CO is proposed:

$$CH_3CHO \rightarrow CH_4 + CO$$

and then, depending on the temperature, reforming of CH_4 and/or WGSR occurs, and, consequently, the extension of both reactions controls the final distribution of products:

$$CH_4 + H_2O \leftrightarrow CO + 3H_2 \quad \Delta H° = 205.8 \text{ kJ mol}^{-1}$$

$$CO + H_2O \leftrightarrow CO_2 + H_2 \quad \Delta H° = -41.1 \text{ kJ mol}^{-1}$$

Concerning rhodium-based catalysts, for Rh/CeO_2-ZrO_2 the direct decomposition of ethanol to CH_4, CO and H_2 have been proposed [29]. This seems to occur via an oxametallacycle intermediate, $-CH_2CH_2O-$, formed from a surface ethoxide species by abstraction of an H from the methyl group by Rh. Then, the reforming of methane and WGS reaction takes place. Other studies, which compare Rh-based catalysts on supports with different ceria content, point to the production of CO_2 as the primary product in the ethanol steam reforming and then its transformation to CO via the WGSR to reach thermodynamic equilibrium [30].

For the steam reforming of ethanol over Rh/Al_2O_3, the mechanism proposed by Cavallaro contemplates the dehydration of ethanol over the support and the dehydrogenation of ethanol over Rh. Then, ethylene is reformed and acetaldehyde is decomposed to CO and CH_4. At 923 K, both reactions are faster than the formation of intermediates [27].

Concerning cobalt-based catalysts, studies of ethanol and acetaldehyde steam reforming employing microcalorimetry and infrared spectroscopy to investigate the adsorption of ethanol and acetaldehyde onto Co/ZnO catalysts, have shown that acetaldehyde transforms to surface acetate species over the fresh catalyst, and these species have been related to the production of H_2 and CO_2 from the steam reforming of acetaldehyde [24,26]. On the other hand, the deactivated catalyst Co/ZnO, which does not show surface acetate species after the acetaldehyde adsorption, promoted the decomposition of acetaldehyde to CO and CH_4 under reforming conditions [26].

Nickel- and Cobalt-Based Catalysts

Many studies of ethanol steam reforming have been carried out over nickel-based catalysts, because these are used effectively on an industrial scale for the steam reforming of natural gas and higher hydrocarbons. Different carriers have been studied [7,31–34], a major concern being the catalyst deactivation by carbon deposition. In this context, basic supports give better results, concerning coke formation, than do acidic carriers.

Two nickel-based systems, Ni/La_2O_3 and Ni/MgO can be highlighted [7,34]. Ni/La_2O_3 showed a good performance in terms of activity, selectivity, and stability [7]. No dehydration products were detected in the course of catalytic tests. Below 573 K, only dehydrogenation of ethanol occurred. The ethanol steam reforming takes place above 673 K. At this temperature, the presence of CO_2 and CH_4 is caused by the WGSR and methanation reaction, respectively:

$$CO + H_2O \leftrightarrow CO_2 + H_2 \quad \Delta H° = -41.1 \text{ kJ mol}^{-1}$$

$$CO + 3H_2 \leftrightarrow CH_4 + H_2O \quad \Delta H° = -205.6 \text{ kJ mol}^{-1}$$

Then, at higher temperatures, the yield of H_2 increases and the selectivity to CO_2 and CH_4 decreases. On the one hand, high temperature does not favor the WGSR. On the other hand, the reforming of CH_4 with H_2O and/or CO_2 (dry reforming of methane) could take place, because these reactions become thermodynamically feasible above 823 K. Moreover, the Ni/La_2O_3 catalyst was seen to be very active for these reforming reactions:

$$CH_4 + H_2O \leftrightarrow CO + 3H_2 \quad \Delta H° = 205.6 \text{ kJ mol}^{-1}$$

$$CH_4 + CO_2 \leftrightarrow 2CO + 2H_2 \quad \Delta H° = 246.9 \text{ kJ mol}^{-1}$$

It is worth mentioning that there is good long-term stability of Ni/La_2O_3 when it is compared with other nickel-supported catalysts. This feature has been attributed to the lack of formation of carbon deposits on its surface, and a model has been proposed for this [7]. A lanthanum oxycarbonate species is formed when CO_2 reacts with a La_2O_x species:

$$La_2O_3 + CO_2 \rightarrow La_2O_2CO_3$$

La$_2$O$_2$CO$_3$, which decorates the nickel particles, removes the surface carbon located at its periphery; the following reaction has been proposed:

$$La_2O_2CO_3 + C \rightarrow La_2O_3 + 2CO$$

The Ni/MgO system has been proposed as appropriate to carry out the steam reforming of bioethanol to supply H$_2$ to MCFC [34]. The addition of alkaline ions produces metal particles of larger size and higher specific activity. The authors claimed that potassium addition stabilizes nickel catalysts by depressing the metal sintering under steam reforming conditions [34]. It has been suggested that the addition of potassium could change the electronic properties of Ni/MgO catalysts by electronic transfer from alkali-oxide moieties to nickel particles, which may depress the Boudouard reaction (2CO $\rightarrow$ CO$_2$ + C) and hydrocarbon decomposition, which can lead to coke formation during steam reforming [31].

On the other hand, several studies have been carried out over Ni–Cu-based catalysts [35–39]. When alumina was used as support [35–37]. Potassium was added to avoid dehydration reactions. The study of catalyst generation as a function of the calcination step and the reducibility of different phases has been analyzed. Copper has been proposed to be responsible for the fast ethanol dehydrogenation to acetaldehyde and nickel for the C–C rupture of acetaldehyde to produce methane and CO. A new aspect was recently introduced when Cu–Ni catalysts supported on SiO$_2$ were considered: the formation of Cu–Ni, alloys which may prevent the deactivation of catalysts by carbon deposits [40].

As stated above, cobalt is also considered an appropriate active phase for the steam reforming of ethanol. An early study on supported cobalt-based catalysts was reported by Haga et al. [41]. This study provided evidence that the support strongly influences the catalyst performance. However, most of the cobalt-based catalysts that were tested (ZrO$_2$-, MgO-, SiO$_2$- and C-supported) showed high yields of methane, probably coming from ethanol or acetaldehyde decomposition or CO hydrogenation, and these reactions were highly suppressed over Co/Al$_2$O$_3$. Several later studies have been devoted to clarifying the role of support and cobalt phases in the behavior of catalysts [21,42–45]. Unsupported or ZnO-supported Co$_3$O$_4$ transforms under steam reforming conditions to small Co particles, which show a high catalytic performance in ethanol steam reforming [43,45]. Using bioethanol-like mixtures (H$_2$O/CH$_3$CH$_2$OH = 13, molar ratio), these catalysts can yield up to 5.5 moles of hydrogen per mole of ethanol reacted at 623 K. The use of relatively low reaction temperature and an excess of H$_2$O makes it possible to obtain almost exclusively CO$_2$ and H$_2$ as reaction products with a low presence of by-products [43,45]. The addition of alkaline metals on Co/ZnO catalysts has been found to have a promoter effect because they stabilize the catalyst by inhibiting coke formation [25].

Noble Metal-Based Catalysts

A review of ethanol reactions over the surface of supported noble metal catalysts has recently been published [46]. The noble metal-based catalysts most widely studied in ethanol steam reforming are those based on Pd, Pt, Ru, and Rh, and their behavior also depends, in this case, on the support. Comparative studies of g-Al_2O_3-supported catalysts showed Rh to be the most active metallic phase [47,48]. Taking into account that ethylene and methane were the main by-products, the performance of different metals in the reformation of ethylene and methane is a key aspect to take into account. Rh turned out to be the most active in ethylene steam reforming, whereas Pd was almost inactive. In this context, the selectivity to ethylene of an Rh/Al_2O_3 catalyst showed a maximum at 973–1023 K, and then dropped at higher temperatures because ethylene reforming took place [47]. To optimize the hydrogen production by steam reforming of bioethanol on Rh/Al_2O_3 catalyst, high temperature and long contact times (high reactor volume/volumetric flow rate ratios) are required [8]. Although Rh is highly active in hydrogenation and its presence may reduce coke formation, under steam reforming conditions the catalyst deactivates by sintering and coke formation. The introduction of a small amount of O_2 (0.4 vol%) has been proposed to reduce coke formation by combustion of carbonaceous species forming during the reaction. However, this combustion could be responsible for the formation of larger metal particles and consequently of the decrease in activity [8].

Several papers have been published in which the reforming of ethanol is carried out over Rh on supports other than alumina, namely, CeO_2, ZrO_2, and derived systems [29,49]. A high yield in H_2 is found for catalysts containing ZrO_2. This is related to the available oxygen on the surface, which participates in the WGS reaction [49].

As for palladium catalysts, it has recently been shown that the steam reforming of ethanol can be effectively carried out over a commercial Pd/g-Al_2O_3 catalyst, which does not produce ethylene as a by-product [23]. At 473–623 K, ethanol is dehydrogenated to acetaldehyde, which is decomposed to CH_4 and CO. Then, at temperatures higher than 733 K, CH_4 can be reformed as a function of the H_2O/ethanol ratio.

Other promising systems for the steam reforming of ethanol could contemplate the concurrence of different catalysts, with an appropriate active phase to catalyze each step of the total process. A two-layer fixed-bed reactor containing Pd/C and a Ni-based catalyst has been proposed to produce CO_x:H_2 mixtures from ethanol steam reforming [50]. Ethanol can be converted to carbon oxides, CH_4 and H_2 over the Pd/C catalyst (608 K). Then, over the second layer containing the Ni-based catalyst, methane can be reformed with steam (923–1073 K) [50].

CATALYSTS FOR AUTOTHERMAL STEAM REFORMING

As mentioned above, one possibility of operation with a more favorable energetic balance is under autothermal steam reforming conditions, which are produced by introducing oxygen in the reaction mixture. As has occurred for hydrocarbons and methanol feedstock [5,51], research is now under way to develop catalysts that control the oxidation process through the combining of catalytic partial oxidation and steam reforming of ethanol. The oxidizing environment reduces the carbon poisoning of the catalyst and could promote the decomposition of intermediate molecules such as ethylene and acetaldehyde. On the other hand, an excess of oxygen leads to a strong reduction of hydrogen as reaction product. In this respect, studies under partial oxidation conditions could contribute to a better knowledge of autothermal ethanol steam reforming. Some studies on catalytic behavior of Ni- [52], Pt- [53] and Ru-based [54] catalysts have recently been reported.

In autothermal conditions, reports concerned the use of Ni and Cu catalysts [38–40,55] and promoted noble metals supported on highly stable carriers, i.e., $Pt-CeO_2-La_2O_3/Al_2O_3$ [13], Rh/CeO_2 [14], Rh/Al_2O_3 [56]. The main role of promoters is related in this case to metal-promoter interactions [13], which affect the adsorption-decomposition of ethanol to CH_4 and CO and their subsequent reforming with steam to produce H_2. Most of the results reported point to the need to operate in a narrow range of water/oxygen/ethanol ratios to achieve 100% ethanol conversion, maximum hydrogen yield and minimum methane and carbon monoxide production.

PERSPECTIVES

As we have mentioned above, one difficulty to be overcome for the practical and extensive use of biomass-derived ethanol as a hydrogen source to fuel-cell systems is supplying the energy needed to distill and/or vaporize the H_2O/ethanol mixtures, and that related to the endothermicity of the steam reforming reaction.

Recently, Dumesic and coworkers have shown that methanol, ethylene glycol, glycerol, and sorbitol can be reformed in the aqueous phase to H_2 and CO_2 at temperatures near 500 K and at pressures between 15–50 bar [57–59]. On the basis of these studies, the reforming of ethanol in the aqueous phase appears as a new approach to be considered for the production of H_2 from ethanol reformation. This process would have several advantages over steam reforming: i) it does not need energy to vaporize alcohol and water before the reaction; ii) the operating temperatures and pressures are suitable for the water–gas shift reaction, so it may be possible to generate hydrogen with low amounts of CO in a single step; and iii) the step of H_2 purification or CO_2 separation is simplified because of the pressure range of the effluent.

Another possibility that merits greater study is the operation under autothermal conditions with ethanol/water/air mixtures. Here, the goal is to maximize the hydrogen yield, while minimizing the total combustion and the formation of by-products and carbon deposits on the catalysts. For both steam reforming and oxidative steam reforming, future research is needed to develop more stable, active, selective, and inexpensive catalytic systems that operate under the required final experimental conditions.

Finally, the integration of the ethanol reformation in an energetically favored total process is also an area, which, still in our day, remains to be completed from a technological point of view.

Efforts in the above-mentioned areas could lead to the practical use of ethanol as H_2 supplier to generate clean electrical power in the not-so-distant future.

REFERENCES

1. Wheeler, C., Jhalani, A., Klein, E.J., Tummala, S. and Schmidt, L.D., The water-gas-shift reaction at short contact times, *J. Catal.*, 233, 191–199, 2004.
2. Schumacher, N., Boisen, A., Dahl, S., Gokhale, A.A., Kandoi, S., Grabow, L.C., Dumesic, J.A., Mavrikakis, M. and Chorkendorff, I., Trends in low-temperature water-gas shift reactivity on transition metals, *J. Catal.*, 229, 265–275, 2005.
3. Comas, J., Laborde, M. and Amadco, N., Thermodynamic analysis of hydrogen production from ethanol using CaO as a CO_2 sorbent, *J. Power Sources*, 138, 61–67, 2004.
4. Semelsberger, T.A., Brown, L.F., Borup, R.L. and Inbody, M.A., Equilibrium products from autothermal processes for generating hydrogen-rich fuel-cell feeds, *Int. J. Hydrogen Energy*, 29, 1047–1064, 2004.
5. Rostrup-Nielsen, J.R., Conversion of hydrocarbons and alcohols for fuel cells, *Phys. Chem. Chem. Phys.*, 3, 283–288, 2001.
6. Rostrup-Nielsen, J.R. and Christiansen, L.J., Internal steam reforming in fuel cells and alkali poisoning, *Appl. Catal. A: General*, 126, 381–390, 1995.
7. Fatsikostas, A.N., Kondarides, D.I. and Verykios, X.E., Production of hydrogen for fuel cells by reformation of biomass-derived ethanol, *Catal. Today*, 75, 145–155, 2002.
8. Cavallaro, S., Chiodo, V., Freni, S., Mondello, N. and Frusteri, F., Performance of Rh/Al_2O_3 catalyst in the steam reforming of ethanol: H_2 production for MCFC, *Appl. Catal. A: General*, 249, 119–128, 2003.
9. Assabumrungrat, S., Pavarajarn, V., Charojrochkul, S. and Laosiripojana, N., Thermodynamic analysis for a solid oxide fuel cell with direct internal reforming fueled by ethanol, *Chem. Eng. Sci.*, 59, 6015–6020, 2004.
10. Wanat, E.C., Venkataraman, K. and Schmidt, L.D., Steam reforming and water-gas shift of ethanol on Rh and Rh-Ce catalysts in a catalytic wall reactor, *Appl. Catal. A: General*, 276, 155–162, 2004.
11. Brown, L.F., A comparative study of fuels for on-board hydrogen production for fuel-cell-powdered automobiles, *Int. J. Hydrogen Energy*, 26, 381–397, 2001.
12. Ioannides, T., Thermodynamic analysis of ethanol processors for fuel cell applications, *J. Power Sources*, 92, 17–25, 2001.
13. Navarro, R.M., Álvarez-Galván, M.C., Sánchez-Sánchez, M.C., Rosa, F. and Fierro, J.L.G., Production of hydrogen by oxidative reforming of ethanol over Pt catalysts supported on Al_2O_3 modified with Ce and La, *Appl. Catal. B: Environmental*, 55, 229–241, 2005.

14. Deluga, G.A., Salge, J.R., Schmidt, L.D. and Verykios, X.E., Renewable hydrogen from ethanol by autothermal reforming, *Science*, 303, 993–997, 2004.

15. Douvartzides, S.L., Coutelieris, F.A. and Tsiakaras, P.E., On the systematic optimization of ethanol fed SOFC-based electricity generating systems in terms of energy and exergy, *J. Power Sources*, 114, 203–212, 2003.

16. Douvartzides, S.L., Coutelieris, F.A., Demin, A.K. and Tsiakaras, P.E., Fuel options for solid oxide fuel cells: a thermodynamic analysis, *AIChE Journal*, 49, 248–257, 2003.

17. Ioannides, T. and Neophytides, S., Efficiency of solid polymer fuel cell operating on ethanol, *J. Power Sources*, 91, 150–156, 2000.

18. Iwasa, N. and Takezawa, N., Reforming of ethanol. Dehydrogenation to ethyl acetate and steam reforming to acetic acid over copper-based catalysts, *Bull. Chem. Soc. Jpn.*, 64, 2619–2623, 1991.

19. Cavallaro, S. and Freni, S., Ethanol steam reforming in a molten carbonate fuel cell. A preliminary kinetic investigation, *Int. J. Hydrogen Energy*, 21, 465–469, 1996.

20. Somorjai, G., *Introduction to Surface Chemistry and Catalysis*, Wiley, New York, 1994.

21. Llorca, J., Ramírez de la Piscina, P., Sales, J. and Homs, N., Direct production of hydrogen from ethanolic aqueous solutions over oxide catalysts, *Chem. Commun.*, 641–642, 2001.

22. Freni, S., Cavallaro, S., Mondello, N., Spadaro, L. and Frusteri, F., Steam reforming of ethanol on Ni/MgO catalysts: H_2 production for MCFC, *J. Power Sources*, 108, 53–57, 2002.

23. Goula, M.A., Kontou, S.K. and Tsiakaras, P.E., Hydrogen production by ethanol steam reforming over a commercial Pd/g-Al$_2$O$_3$ catalyst, *Appl. Catal. B: Environmental*, 49, 135–144, 2004.

24. Llorca, J., Homs, N. and Ramírez de la Piscina, P., In situ DRIFT-mass spectrometry study of the ethanol steam-reforming reaction over carbonyl-derived Co/ZnO catalysts, *J. Catal.*, 227, 556–560, 2004.

25. Llorca, J., Homs, N., Sales, J., Fierro, J.L.G. and Ramírez de la Piscina, P., Effect of sodium addition on the performance of Co-ZnO-based catalysts for hydrogen production from bioethanol, *J. Catal.*, 222, 470–480, 2004.

26. Guil, J.M., Homs, N., Llorca, J. and Ramírez de la Piscina, P., Microcalorimetric and infrared studies of ethanol and acetaldehyde adsorption to investigate the ethanol steam reforming on supported cobalt catalysts, *J. Phys. Chem. B*, 109, 10813–10819, 2005.

27. Cavallaro, S., Ethanol steam reforming on Rh/Al$_2$O$_3$ catalysts, *Energy & Fuels*, 14, 1195–1199, 2000.

28. Frusteri, F., Freni, S., Spadaro, L., Chiodo, V., Bonura, G., Donato, S. and Cavallaro, S., H_2 production for MC fuel cell by steam reforming of ethanol over MgO supported Pd, Rh, Ni and Co catalysts, *Catal. Commun.*, 5, 611–615, 2004.

29. Diagne, C.; Idriss, H.; Kiennemann, A. Hydrogen production by ethanol reforming over Rh/CeO$_2$-ZrO$_2$ catalysts, *Catal. Commun.*, 3, 565–571, 2002.

30. Auprêtre, F., Descorme, C. and Duprez, D., Bio-ethanol catalytic steam reforming over supported metal catalysts, *Catal. Commun.*, 3, 263–267, 2002.

31. Frusteri, F., Freni, S., Chiodo, V., Spadaro, L., Bonura, G. and Cavallaro, S., Potassium improved stability of Ni/MgO in the steam reforming of ethanol for the production of hydrogen for MCFC, *J. Power Sources* 132, 139–144, 2004.

32. Fatsikostas, A.N., Kondarides, D.I. and Verykios, X.E., Steam reforming of bio-mass-derived ethanol for the production of hydrogen for fuel cell applications, *Chem. Commun.*, 851–852, 2001.

33. Fatsikostas, A.N. and Verykios, X.E., Reaction network of steam reforming of ethanol over Ni-based catalysts, *J. Catal.*, 225, 439–452, 2004.

34. Frusteri, F., Freni, S., Chiodo, V., Spadaro, L., Di Blasi, O., Bonura, G. and Cavallaro, S., Steam reforming of bio-ethanol on alkali-doped Ni/MgO catalysts: hydrogen production for MC fuel cell, *Appl. Catal. A: General*, 270, 1–7, 2004.

35. Mariño, F., Baronetti, G., Jobbagy, M. and Laborde, M., Cu-Ni-K/g -Al$_2$O$_3$ supported catalysts for ethanol steam reforming. Formation of hydrotalcite-type compounds as a result of metal-support interaction, *Appl. Catal. A: General*, 238, 41–54, 2003.

36. Mariño, F.; Boveri, M.; Baronetti, G.; Laborde, M. Hydrogen production from steam reforming of bioethanol using Cu/Ni/K/g-Al$_2$O$_3$ catalysts. Effect of Ni. *Int. J. Hydrogen Energy*, 26, 665–668, 2001.

37. Mariño, F.J., Cerrella, E.G., Duhalde, S., Jobbagy, M. and Laborde, M.A., Hydrogen from steam reforming of ethanol. Characterization and performance of copper-nickel supported catalysts, *Int. J. Hydrogen Energy*, 23, 1095–1101, 1998.

38. Klouz, V., Fierro, V., Denton, P., Katz, H., Lisse, J.P., Bouvot-Mauduit, S. and Mirodatos, C., Ethanol reforming for hydrogen production in a hybrid electric vehicle: process optimisation, *J. Power Sources*, 105, 26–34, 2002.

39. Fierro, V., Klouz, V., Akdim, O. and Mirodatos, C., Oxidative reforming of biomass derived ethanol for hydrogen production in fuel cell applications, *Catal. Today*, 75, 141–144, 2002.

40. Fierro, V., Akdim, O. and Mirodatos, C., On-board hydrogen production in a hybrid electric vehicle by bio-ethanol oxidative steam reforming over Ni and noble metal based catalysts, *Green Chemistry* 5, 20–24, 2003.

41. Haga, F., Nakajima, T., Miya, H. and Mishima, S., Catalytic properties of supported cobalt catalysts for steam reforming of ethanol, *Catal. Lett.*, 48, 223–227, 1997.

42. Llorca, J., Homs, N., Sales, J. and Ramírez de la Piscina, P., Efficient production of hydrogen over supported cobalt catalysts from ethanol steam reforming, *J. Catal.*, 209, 306–317, 2002.

43. Llorca, J., Ramírez de la Piscina, P., Dalmon, J.A. and Homs, N., Transformation of Co$_3$O$_4$ during ethanol steam-reforming activation process for hydrogen production, *Chem. Mater.*, 16, 3573–3578, 2004.

44. Llorca, J., Dalmon, J.-A., Ramírez de la Piscina, P. and Homs, N., *In situ* magnetic characterisation of supported cobalt catalysts under steam-reforming of ethanol, *Appl. Catal. A: General*, 243, 261–269, 2003.

45. Llorca, J., Ramírez de la Piscina, P., Dalmon, J.-A., Sales, J. and Homs, N., CO-free hydrogen from steam-reforming of bioethanol over ZnO-supported cobalt catalysts, Effect of the metallic precursor, *Appl. Catal. B: Environmental*, 43, 355–369, 2003.

46. Idriss, H., Ethanol reactions over surfaces of noble metal/cerium oxide catalysts, *Platinum Metals Rev.*, 48, 105–115, 2004.

47. Liguras, D.K., Kondarides, D.I. and Verykios, X.E., Production of hydrogen for fuel cells by steam reforming of ethanol over supported noble metal catalysts, *Appl. Catal. B: Environmental*, 43, 345–354, 2003.

48. Breen, J.P., Burch, R. and Coleman, H.M., Metal-catalysed steam reforming of ethanol in the production of hydrogen for fuel cell applications, *Appl. Catal. B: Environmental*, 39, 65–74, 2002.
49. Diagne, C., Idriss, H., Pearson, K., Gómez-García, M.A. and Kiennemann, A., Efficient hydrogen production by ethanol reforming over Rh catalysts. Effect of addition of Zr on CeO_2 for the oxidation of CO to CO_2, *CR Chimie*, 7, 617–622, 2004.
50. Galvita, V.V., Semin, G.L. Belyaev, V.D., Semikolenov, V.A., Tsiakaras, P. and Sobyanin, V.A., Synthesis gas production by steam reforming of ethanol, *Appl. Catal. A: General*, 220, 123–127, 2001.
51. Peña, M.A., Gomez, J.P. and Fierro, J.L.G., New catalytic routes for syngas and hydrogen production, *Appl. Catal. A: General*, 144, 7–57, 1996.
52. Liguras, D.K., Goundani, K. and Verykios, X.E., Production of hydrogen for fuel cells by catalytic partial oxidation over structured Ni catalysts, *J. Power Sources*, 130, 30–37, 2004.
53. Mattos, L.V. and Noronha, F.B., Partial oxidation of ethanol on supported Pt catalysts, *J. Power Sources*, 2005, in press.
54. Liguras, D.K., Goundani, K. and Verykios, X.E., Production of hydrogen for fuel cells by catalytic partial oxidation of ethanol over structured Ru catalysts, *Int. J. Hydrogen Energy*, 29, 419–427, 2004.
55. Velu, S., Satoh, N., Gopinath, Ch.S. and Suzuki, K., Oxidative reforming of bio-ethanol over CuNiZnAl mixed oxide catalysts for hydrogen production, *Catal. Lett.* 82, 145–152, 2002.
56. Cavallaro, S., Chiodo, V., Vita, A. and Freni, S., Hydrogen production by auto-thermal reforming of ethanol on Rh/Al_2O_3 catalyst, *J. Power Sources*, 123, 10–16, 2003.
57. Cortright, R.D., Davda, R.R. and Dumesic, J.A., Hydrogen from catalytic reforming of biomass-derived hydrocarbons in liquid water, *Nature*, 418, 964–967, 2002.
58. Huber, G.W., Shabaker, J.W. and Dumesic, J.A., Raney Ni-Sn catalyst for H_2 production from biomass-derived hydrocarbons, *Science*, 300, 2075–2077, 2003.
59. Davda, R.R., Shabaker, J.W., Huber, G.W., Cortright, R.D. and Dumesic, J.A., A review of catalytic issues and process conditions for renewable hydrogen and alkanes by aqueous-phase reforming of oxygenated hydrocarbons over supported metal catalysts, *Appl. Catal. B: Environmental*, 56, 171–186, 2005.

14 Ethanol from Bakery Waste: The Great Provider for Aquaponics?

Robert Haber
One Accord Food Pantry, Inc.
Troy, New York

CONTENTS

INTRODUCTION

The evolution of this project took a period of over 20 years. Originating in the pre-Reagan era of what we once thought were high gasoline prices, the concept was to simply make ethanol for electric or transportation use and feed the by-products to pigs and chickens. This still left remaining waste to manage, however. With the change of politics and policies, all federal grants for alcohol research were cancelled. As a result, the concept went unfulfilled for 20 years. But the world is a different place now. Today, the concept has evolved to design and implement a zero-discharge, closed, recirculating, environmentally isolated system, which produces microelement-enhanced, high-quality protein food using municipal solid waste as a source for nonpetroleum power generation. After having been the 13-year director of a rural food pantry, which met the emergency food needs of over 40,000 rural needy a year (half of whom were children), I enjoyed a unique perspective of the massive amounts of food waste that are discarded daily, especially breads and bakery sweets. As a resource for this project, the huge quantity of useable bakery waste was staggering and dictated the type of fuel to be made. At issue for the project were not only the need to generate heat and electricity, but also the need to have an ingredient base for on-site manufactured fish feed. These three expenses (heat, electric, and feed) comprise the bulk of all operating expenses associated with the long-term success or failure of the project. Reducing or eliminating these expenses would then enhance the economic viability and potential success of the project. Of critical importance were the ingredients for the feed, since it was the only nutrient input into the system for both fish and plants. Only one type of fuel met all three needs — ethanol — and in particular, ethanol from bakery waste. Additionally, and for the purposes of this project, were the by-products of fermentation (carbon dioxide and DDGS (distillers dried grains and solubles)) and combustion (carbon dioxide and water vapor). The following is an in-depth description of the project.

THE PROJECT — PHASE 1

The first phase of this project is composed of the following subsystems:

1. Alcohol fuel.
2. Solid MSW fuel — wood and cardboard.
3. Aquaponics.
4. Fish feed formulation.
5. Fish hatchery.
6. Energy plantation.
7. Compost.
8. Processing.
9. Technology transfer — website.

ALCOHOL FUEL

One greenhouse (50' × 156') will be used to house the alcohol fuel production equipment and fish feed equipment. It will be located separately from the aquaponic greenhouses. Site preparation has been completed for this greenhouse, as well as water supply and electric transmission lines. The water supply is from a developed artesian well, which will have its capacity expanded. The project will use bakery waste as the feedstock for alcohol since it is so plentiful locally. After packaging is removed, the production of fuel will follow these steps: A) the bakery waste is passed through a standard garden chipper-shredder, B) the shredded bakery waste is mixed with hot water in a tank, C) a liquefication enzyme is added and the mixture is boiled for about 20 minutes — this enzyme prevents the slurry from jelling, D) the mash is cooled to 140°F and a saccharification enzyme is added. It is held at this temperature for another 20 minutes. The saccharification enzyme converts starch to sugar, E) once starch conversion is complete, the mash is cooled to 90°F, adjusted for optimum yeast activity to occur (Ph and brix), and distillers yeast is added. A vapor lock is installed to eliminate contamination of the ferment by airborne putrescent bacteria. CO_2 is captured at this stage, F) when yeast activity stops — from 3–5 days — the ferment is filtered to remove solids, and G) the clear liquid beer is now ready for distillation. Once distilled and denatured, it is ready for the microturbine to generate electricity and heat for the greenhouses. The solids that are generated will be mixed with other components, pelletized, and used as the base for fish feed. Packaging is sorted, compressed, and sold as scrap, or used as direct combustibles.

MUNICIPAL SOLID WASTE (MSW) FUEL — WOOD AND CARDBOARD

Carbon-Cycle Neutral

Two types of potential fuels (cardboard and wood) are presently a major part of the Municipal Solid Waste (MSW) stream, and comprise approximately 45.8%

of all MSW composition (1). These green energy fuels are organic in origin and renewable in nature; therefore, they are carbon-cycle neutral, i.e., since they are composed of former living plant tissue, when used as fuel for heat or converted to alcohol fuel for electric generation, they do not add to greenhouse gasses or carcinogens to the environment. The carbon dioxide that is emitted is used by the next generation of growing plants to store energy; therefore, the carbon cycle is stabilized through their use. Petroleum fuels add carbon dioxide to the cycle, which has been sequestered and removed from the cycle for millennia.

Procedure

Wood and cardboard will be removed from the waste stream at the source. The project will utilize nonrecyclable cardboard and wood as direct combustible fuel. Compressed, baled, and placed on a pallet for ease of handling, nonrecyclable cardboard and other solid combustibles will be fed directly into a specially designed outdoor furnace that generates 1,000,000 BTUs per hour. Ash generated is rich in boron, an element normally deficient in New England soils. It will be incorporated into compost.

Potential Savings

As an example of the potential savings to be realized from utilizing these two forms of MSW, the BTUs in just one 500-pound bale of cardboard (at 8200 BTUs/lb) are equal to the BTUs in 29.7 gallons of #2 home heating oil (at 138,000 BTUs/gallon). Given the fact that one barrel (42 gallons) of crude oil only yields 9.2 gallons of home heating oil or diesel (2), the savings from using one ton of cardboard is the equivalent heating oil yield from 12.92 barrels of crude oil. There are approximately 7500 BTUs in one pound of wood. Given the above information, the project will be a net reducer of MSW that will contribute to the overall reduction of landfill mass.

AQUAPONICS

Aquaponics is the joining together of two food-producing systems, aquaculture (food fish farming) and hydroponics (soilless vegetable farming). When these two systems are joined, they form a symbiotic relationship with each other (each benefits from the other). Fish breathe in the same water in which they eliminate, creating an overabundance of ammonia waste and a deficiency of oxygen. If the oxygen is not replaced and the ammonia waste not removed, the fish will die. Using the effluent from the fish tanks to grow plants does two things: first, the plants remove the nitrogenous wastes from the water through their roots and use it for growth, second, the clean water is then oxygenated and returned to the fish tank. The only nutrient input into the system is fish feed. The dimensions for the concrete grow-out tank are 4' high, 20' wide and 60' long. This tank has the capability to produce 1 metric ton of fish weekly (2200 pounds), and will carry an average of 18,000 pounds of fish at all stages of growth. The tank is a modified raceway, which is folded back on itself, i.e., it is a 10' wide raceway folded back, which now makes it 20' wide with a divider in the middle. The water flow is straight through. The return from the grow beds enters on the right side of the tank and flows all the way around to exit on the left side, carrying solid wastes with it. A 2" recess in the tank floor on the left side allows the solids to accumulate and be pumped to the grow beds. Insulated plumbing will connect the tanks to the aquaponic grow beds in other parts of greenhouse. These beds use pea gravel as a growing medium and measure 4' × 8' × 1', and are elevated to hip height, eliminating stooping. These beds are required to provide adequate biofiltration for the fish tank and will provide approximately 9888 square feet of plant growing area. The surface of the gravel will provide growing space for nitrifying bacteria, which convert the fish wastewater to a useable form for the growing plants to absorb. The growing beds, therefore, act as a biofilter to cleanse the water for the fish and the fish provide nutrients for the plants, which are so stimulating to the plants that days-to-maturity are often reduced by 1/3 to 1/2. The grow beds are flooded to 1" beneath the top surface of the gravel every hour for 3–5 minutes. The water is then drained by gravity into the sump tanks and pumped back into the fish tank. Project research has not discovered any explanation for this astounding growth rate, so it remains a mystery. However, empirical evidence is very real as observed by the effect on field-grown red raspberries (the reader is invited to see picture documentation on pantry homepage at: www.oneaccordfoodpantry.org and specific weeds, i.e., Queen Anne's Lace — or wild carrot, nettles, and goldenrod — which attained heights of approximately 9'. The effect was also evident on strawberries which reached hipheight and had stems as thick as one's little finger.

Fish Produced

The type of fish the project will use are tilapia. They are a very forgiving fish, adaptable to a wide range of environments. They are fecund (have lots of babies) and grow from birth to marketable size in 6 months. They are efficient users of feed (5 pounds of feed = 1 pound of fish). By contrast, beef requires 19 pounds of feed to produce 1 pound of gain. Tilapia are the answer to a hungry world. The flesh is mild and flavorful. They are mouth breeders (the mother carries the eggs in her mouth until they hatch) and a tropical fish that require warm water, so there is no danger of their "escape" from cultivation and contamination of local waterways.

Auto-Feeders

The fish finish tank will be equipped with auto-feeders, which hold a reserve of feed and are activated by hungry fish bumping a rod that projects down into the water. When activated, it releases a small amount of pellets to the fish below. The fish always have a supply of feed before them that is released on their demand. As a result, the feed is always fresh and overfeeding is reduced thereby enhancing the quality of tank water. Water quality drops when uneaten feed is present on the bottom of the tank. This eliminates the need for 5–6 hand feedings throughout the day. Labor then is only required to assure that the auto-feeders are kept full. The fish are trained in the hatchery to receive food in this manner.

Roof of the Structure

The roof of the aquaponics greenhouse is designed to capture rainwater and to melt and capture snow. Water is then transported through pipes to a central water-storage tank. Since this geographic area is subject to periodic droughts, the need to keep a supply of water is essential during the summer months.

Floor of the Structure

The floor of the aquaponics greenhouse is 6" thick concrete fitted with special plumbing that allows for the passage of hot water during the winter and cold water during the summer. It is connected to both the outdoor furnace and the recouperators on the microturbines. It has 3" of special insulation below it. Sump tanks, which gather flood water from the growbeds, are beneath the floor and accessible through a manhole in the top. It is then pumped back into the fish tank.

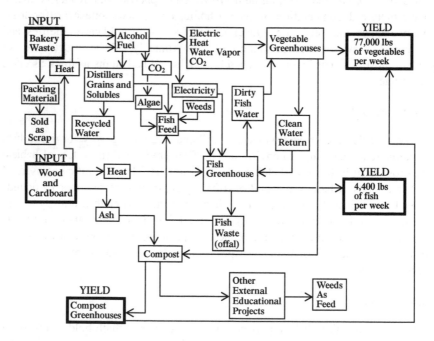

FUEL   FLOW   CHART

FIGURE 14.1 Flow chart of system.

Balanced Diet

Many types of plants can be grown in this system, if all of their cultural requirements are met (light, heat, nutrients, and water). There are no weeds, so each plant can reach its full potential. The beds are comfortable to work with since they are hip height and require no stooping. As such, the project can potentially be used to provide a complete diet. Initially, the project will produce lettuce, broccoli, tomatoes, parsley, oregano, basil, cucumbers, peppers, onions, swiss chard, and spinach.

Efficient Use of Water

Whereas the aquaponics system is water-based, the system as a whole is very efficient in the use of water to distribute nutrients. There is a small amount of evaporation from the fish tanks and growbeds, but the majority of water used goes into plant and fish growth, i.e., the only water actually removed from the system is in the form of plant material and fish flesh.

Power Generation

The aquaponics greenhouse will have microturbines placed inside during the winter to generate electricity. A microturbine is an aircraft engine that has been miniaturized and turns an alternator. The turbines are inside because the exhaust from burning the alcohol fuel is composed of carbon dioxide and water vapor — at a temperature of around 1300°F. An 80-kw turbine will generate nearly 900,000 BTUs per hour in its exhaust. The carbon dioxide "fertilizes" the air for the plants, increasing growth potential. The plants in turn release oxygen into the air. Each 30' bay of the greenhouse will require approximately 40 kw of power for lighting during the winter. The lighting is essential even on sunny days due to the low angle of incidence of insolation (incoming solar radiation). The sunlight bounces off the roof of the structure and does not penetrate in enough candlepower to benefit the plants. Without lights, there is a very real danger of nitrogen poisoning in green leafy vegetables grown inside. Heat and carbon dioxide levels will be monitored continuously. Should either exceed acceptable limits, the exhaust from the microturbines will be switched to a recouperator — gathering excess heat — and then outside. The recouperator stores excess heat in the thermal mass of the floor.

Potential Yield

Published results from researchers state that for every pound of fish produced, 50–75 pounds of vegetables are produced (3). Our goal is 35 pounds of vegetables, and each tank has the ability to produce 2200 pounds of fish a week at full capacity (we have room for 2–3 tanks). The project can be duplicated nearly anywhere in the world, and requires very little room. The yield from the aquaponics system is enormous. Square foot for square foot, the yield is 15 times that of traditional soil farming, given the same period of time.

Smaller Family-Sized Unit

The project will convert a smaller $17' \times 48'$ greenhouse to a family-sized unit by scaling down the commercial unit. It will contain a smaller tank capable of producing 200 pounds of fish and enough growbed room to accommodate the waste from the fish. Additional growing area will be used for the growth of root crops in compost. The smaller unit will be able to meet the annual vegetable and fish needs for a family of 6. The technology will differ in this pilot project by using whole kernel corn to heat the radiant floor, and used vegetable oil (UVO or WVO) to directly fuel a diesel-powered electric generator.

Bacteria Production

The project will generate a continuous supply of nitrifying bacteria for addition to the system on a weekly basis. Bacteria, which are an integral and critical part of the system, cannot be shipped during the winter since cold temperatures will kill them. The pantry has grown bacteria previously for its aquaculture project. Weekly additions of balanced bacteria to the system will guarantee a large and healthy population of nitrifying bacteria within the system. Lack of bacteria will result in the accumulation of lethal concentrations of ammonia and nitrites, which can kill fish in a matter of hours. The bacteria must be in balance, i.e., equally strong populations. Since the two types of nitrifying bacteria grow at extremely different rates, it is essential to allow enough time for the two populations to equalize to similar population density. Determining population density is through default. Monitoring culture medium for ammonia, nitrite, and nitrate gives a very accurate indication of bacterial growth. All that is required for monitoring is a water quality test kit.

Fish Feed Formulation

An important secondary benefit of generating alcohol fuel from bakery waste is that the residual solids remaining after processing are largely the protein content of the bakery waste in a concentrated form. Also, waste from fish processing can be dehydrated and substituted for marine, or salt water fish meal, since the protein signature is identical to the fish being raised. Marine fish meal is becoming scarce and expensive due to the overharvesting of native fisheries. Feed components are not a minor consideration in any type of aquaculture project, since feed costs can constitute 35–90% of all production costs. When each tank is at full capacity, feed requirements will be about 900 pounds per day. Waste not appropriate for feed will be composted. Some of this compost will be used during the summer for the outdoor cultivation of common yet specific garden weeds that are rich in nutrients for feeding fish. The carbon dioxide captured during fermentation will be used in a "photobioreactor" for the growth of blue-green algae, which will also be incorporated into fish feed. This algae will also be used to seed the Greenwater System for specific species of fish we intend to raise. Equipment for the processing of fish feed will be situated in about 1/4 of the alcohol fuel greenhouse. Drying will be accomplished through the use of commercial dehydrators and formulation through the use of a hammermill, mixer, and pelletizer. It is the goal of the project to produce all the components of fish feed from plant sources. The only feed component input outside the project is kelp meal, which supplies vital microelements.

FISH HATCHERY AND SEEDLING GREENHOUSE

Hatchery

Tilapia are a tropical fish and require warm water. As a result, shipping fingerlings during the winter is extremely risky. Therefore, the project will produce all needed fingerlings on-site. The project will require from 1100 to 3300 fingerlings per week. It can also act as a regional resource for others that require fingerlings as well. The bulk of the equipment needed to breed these fingerlings is already in place. Whereas the hatchery is somewhat labor intensive, it does guarantee a continuous supply of fingerlings. The males are kept in tanks by themselves and the females are brought to them to mate. Once mating has stopped and the female has a mouth full of eggs, the males are removed and the female is left alone and undisturbed to hatch her young (about 72 hours). After this time, the hatchlings will remain in the mother's mouth for another 3–7 days, taking short excursions outside to feed on minute particulates and then dart back inside. At a time determined by the mother, she spits them all out and she will accept feed again. When this occurs, she is removed to an isolation tank and full fed until she regains weight. She is then ready to breed again. The particular variety of tilapia to be used is patented and produces all males (males grow 40% faster than females), which are a bright red in color. They are a very forgiving fish, adaptable to many different cultural conditions and, from past experience with them, quite easy to breed and raise. Presently, there is enough room for future expansion to other species, i.e., giant Australian Red Claw crawfish (*Cherax quadricarinatus* — a lobster-sized crawfish that lives in fresh water), giant freshwater prawns (*macrobraccium Rosenbergii*) which grow to one pound, and a variety of giant sunfish (bream), which grows to a weight of 5 pounds. All these species, except for sunfish, are tropical or subtropical and must be bred on-site to have a year-long supply. The sunfish will remain isolated from the environment, since interbreeding with native varieties will dilute their genetic uniqueness, resulting in much smaller, stunted fish. The hatchery will be connected to its own smaller aquaponics greenhouse, which will provide biofiltration and a use for the waste generated from the hatchery. It will also add to the weekly harvest of vegetables. Breeding fish on-site will also enable the project to add genetics to its teaching curriculum.

Greenwater System

Both the prawns and crawfish go through a critical juvenile phase in their development. At this time, they become cannibalistic and will feed on each other. Molting is frequent, on the order of every 3–5 days, since the growth rate is so very rapid during this stage of development. When they molt, they remain motionless until the outer skin hardens into a new shell. But this immobility is a signal to other juveniles nearby that they are vulnerable. Before the shell hardens, they can literally be torn to pieces. In a clearwater system, even with adequate hiding places, there is considerable loss. This is because they can see each other, and their appetites are so great at this time, they are almost continually hungry. To address both of these causes for loss, the project will utilize a greenwater system for this stage of development. Greenwater is simply clearwater that has so much algae growing in it, that it becomes opaque and vision is limited to less than one inch, which provides considerable concealment to molting individuals. It looks much like pea soup. The project will deliberately seed the greenwater system with spirulina, a very nutritious form of algae, high in protein and essential amino acids in addition to vitamins, minerals, and microelements. Between feedings, it makes an excellent snack for juveniles. Even though both of these species are cannibalistic in the juvenile phase, between feedings they will forage vigorously on algae, which are all around them. The algae then becomes a continuous supply of high-quality feed — much the same as pasture for ruminants. There is no biofiltration used in this rearing system, rather simple oxygenation. Spirulina act as biofilters by digesting spent feed and wastes from the juveniles for their growth and reproduction. Algae also add oxygen to the water, and remove carbon dioxide. Once past the juvenile stage, these species lose their cannibalistic tendencies and will not return to them unless they are overcrowded and underfed.

Seedling Greenhouse

The seedling greenhouse is a smaller structure designed to provide replacement plants for those harvested in the aquaponics greenhouse. Very shortly after sprouting, the young plants are switched to being irrigated with fish tank effluent. This provides the maximum growth potential in the shortest amount of time. This part of the project will be an extremely rich learning environment.

ENERGY PLANTATION

The majority of the energy crops grown will be hybrid poplar and the project will utilize whole tree technology in the harvest and use of the trees. This means the entire tree is cut during the winter and processed into chips. The stump does not die — the root system is well established by this time — but rather the stump sprouts new shoots, which are then trimmed to one central leader. Because of the extensive root system, the tree reaches its original size again in 3 years rather than 5 and is ready for harvest once more. At least one year will be needed to assess the potential of the property, but in particular to plant cover crops and build soil fertility before planting energy crops. Given these restrictions, it will take 5 years from planting before the project is able to harvest a first crop of trees. Also, the project will be experimenting with salix (willow), alfalfa, and rapeseed (canola) crops during this 5–6 year lag in time. The project will also establish fish grow-out tanks on this land. However, the effluent will be used to irrigate outdoor energy crops during the summer and only produce vegetables during the fall, winter, and spring. The heart of this phase of the project is the gasification unit, which converts crop residues, stems from hay, and wood chips into synthesis gas. Feedstock is placed in the unit, which heats it to about 1500 degrees. The heat chamber is without oxygen so the feedstock will not burn, but rather give off a gas that is then filtered, cooled, and stored. This fuel cannot only provide heat and electricity, but also used directly in an internal combustion engine. The unit can also produce alcohol fuel and diesel fuel, #2 home heating fuel, and accommodate rubber tires as a feedstock. Excess heat from the operation of the gasifier will be used to dehydrate certain components for fish feed including the protein remains (DDGS — distillers dried grains and solubles) from the alcohol fuel project, vegetable waste and fish meal from the aquaponics project and other experimental feeds such as duckweed.

COMPOST

Through normal daily operations, the project will generate a large quantity of high-nutrient-level organic matter, i.e., vegetable trimmings, ash, leaves, grass clippings, and fish offal not suitable for feed. This material will be composted in specially designed containers that conserve nutrients. It will then be tested and adjusted for nutrient balance and ph, and used in separate greenhouses for the production of root crops. Fish tank effluent will be used to irrigate these crops as well. Since the effluent will not return to the tanks in this type of system, it will be replaced with fresh water. However, the plant growth stimulating effect of the fish effluent will still be in force, resulting in vigorous growth of the plants.

PROCESSING

The fish sold will be processed into fillets to increase consumer sales appeal, although the project has potential customers presently who wish to purchase live fish. The project will process the fish in a specially designed portable unit. Processing will be done by special equipment that strips off all the scales from the fish and automatically fillets them. The equipment is very fast, using no more than a few seconds to process each fish. All waste from this processing is saved and made into fish meal, another very important feed ingredient making the project just that much more efficient. The need for marine, or salt water, fish meal is thereby eliminated and the protein signature of the meal is identical to the requirements of the fish being raised.

Technology Transfer — Website

The Food Pantry has established a Web presence. During construction it will be possible to tape record and follow along step by step and document just how the greenhouses were sited, erected, equipped, and operated. As such, this will be a soup-to-nuts type of educational experience, giving students an in-depth picture of the entire project. The Website will be upgraded to an interactive site with periodic live teaching segments — broadcast directly from inside any of the outbuildings live over the Internet through streaming video — which will enable viewers to ask questions in real time. The segments will be archived for downloading as a reference and offered for sale in CD format for a modest fee. In this way, the project can offer 1/2-hour teaching segments on every aspect of the system and from every location within the system through the use of a Web cam. Land lines will connect all the external buildings to project offices and computers. Everything then can be broadcast live over the Internet. A certain amount of upgrading will be required to make the offices suitable for this purpose.

The project will be promoting the use of this system worldwide, and modified to meet almost any climatic conditions. Phase 2 of this project will address the adaptation of this system to differing worldwide climatic needs, and other sources of renewable and sustainable fuels to power the project according to those needs. Through the Web page and e-mail, the project would be able to act as a resource to anyone anywhere in the world. Upon completion, it will be a powerful learning tool.

THE PROJECT — PHASE 2

Whereas the first phase of the project is primarily passive, i.e., simply recapturing waste biomass, the second phase will actively demonstrate the culture, growth, harvest, and use of renewable biomass fuels — biomass here being defined as plants specifically grown for their fuel value. The project will also be conducting research into the production of oilseed crops specific to this geographic region — these crops can be directly converted into biodiesel, and used for either electric generation or transportation. Crops specifically grown for gasification include hybrid poplar, salix (willow), and alfalfa. Crop by-products that can be gasified include corn cobs, corn stover, sawdust, wood chips, and chipped brush. Non-crop items include rubber tires, plastic, and construction debris. Oilseed crops include canola (rapeseed), soybeans, and corn. All energy crops will be irrigated with fish effluent with controls in place to determine the beneficial effect this waste product has on field-grown plants.

Abundance of Biomass

To give some idea of the abundance of biomass, the energy content of all biomass fuels available today would produce an estimated 2740 Quads (one Quad equals 1,000,000,000,000,000 BTUs) (4). Swedish physicist John Holmberg claims we have no energy crisis. He believes that since human society's energy use is only about 1/13,000 of our daily solar income, the simple solution to the "crisis" is to harvest the abundance (5). That harvest is carried out day by day through photosynthesis, and stored in the form of biomass. This project then will unite all of its other energy-dependent educational projects into a cohesive whole by demonstrating the advantages of the local production of multiple-use biomass fuels.

SAVING FAMILY FARMS

The project has specifically chosen an agricultural application in the hopes that some of this information may be used by operating farms as a means to create additional on-farm income or that it may encourage others to enter farming. Approximately 450 farms per year go out of business in New York State due to high energy costs and economic failure (6). With no remaining income, many farms fall prey to land developers, resulting in the loss of irreplaceable farmland forever.

FOOD IMPORTED INTO THE NORTHEAST

It is extremely important to remember that even during the summer, when farms in the Northeast are at peak production, the Northeast still imports 95% of its food. During the winter, this number increases to 98% (7). As a result, 95–98% of all money spent on food in the Northeast leaves the Northeast, creating a tremendous cash flow out of region. At the same time, this represents a huge market that is virtually untapped by local growers. These local food producers have the advantage of not having to transport their product into the area. Therefore, promoting local production of food will save on the associated costs of long-distance hauling of food and the fuel associated with transportation. Environmental impact will also be reduced as a result.

RESERVE FOOD SUPPLY

Additionally, the reserve food supply for the entire Northeast (i.e., all the food on supermarket shelves and in warehouses) will supply the food needs of its inhabitants for only 3 days. This constitutes a terrible vulnerability for the entire region, should anything such as terrorism interrupt this constant influx of food. This project, which promotes the decentralization of fuel production, encourages the expansion of local food production, and may potentially save family farms, will address all of these community problems.

The second phase of this project is composed of the following subsystems:

1. Energy plantation — gasification
2. Growth of fish feed from plants
3. Compost-based aquaponic greenhouses
4. Duplication in inner cities

Energy Plantation

The plantation will consist of hybrid poplar trees (at 600 trees per acre) planted in a "lawn" of Dutch white clover. During the first year of growth, the young trees can tolerate no competition from weeds and require irrigation in order to become established. The project will mow the "lawn" with a bagging commercial mower to remove the clippings, which will then be dehydrated and used as a base for fish feed, or used to produce compost. Colonies of honeybees will be established to forage on the clover, since these plants produce an excellent water-white honey. As a result, the project will harvest 3 different crops from the same piece of land. The trees are ready to harvest the 5th year and will regenerate to harvestable size every 3 years thereafter. One pound of hardwood chips generates approximately 7500 BTUs of energy and poplar contains about 60% of this value — or about 4500 BTUs per pound. This reduction in heat value is more than offset by the rapid growth and regeneration. Poplar also absorbs and stores more carbon dioxide during growth than the wood gives off when used as a fuel source. Harvest will commence after leaf drop and when the ground is frozen to reduce turf damage. The project will use a feller/buncher to cut the trees, which are then placed in windrows. A self-propelled chipper reduces the entire tree into 2" or less sized chips — this size being optimal for the gasifier. Chips can then be stored in a silo that self-feeds directly into the gasifier. After this point the entire system is automatic and needs no operator — all that is required is a continuous supply of feedstock (chips). At the heart of this phase is a gasification unit, which heats and thermally degrades the biomass in a chamber devoid of oxygen. The gas generated from this heating (pyrolysis) is extracted, filtered, cooled, and stored. It contains approximately 50% of the BTUs in natural gas and can be used for electric, heat generation, or transportation. The difference here is that gasification can utilize so many other different feedstocks that are unsuitable for the first phase. There is no exhaust from this process, since there is no active combustion, as a result, there is no pollution. The "exhaust" is actually the gas produced. The gasification unit can also be configured to produce methanol (wood alcohol), which is used in the conversion (transesterification) of raw and used vegetable oil into biodiesel. It can also produce dimethylether ("DME"), a clean-burning replacement for diesel fuel, or #2 home heating fuel — depending on the configuration. Solids and nutrient ash from wood biomass pyrolysis can be incorporated into compost as a bulking/nutrient agent. Inert solids remaining from plastic and tire pyrolysis can be used for making asphalt paving products or concrete blocks. The only feedstock under consideration that is not carbon-cycle neutral is plastic.

Growth of Fish Feed from Plant Sources

Excess heat generated from gasification can be captured and used for process heat in any application. Since there will be no edible residuals from gasification, fish feed will have to be produced in another manner. To this end, we will conduct production and use trials with feeds ranging from alfalfa leaf, small grains, soy, corn, legume lawn clippings, algae (spirulina), and duckweed. The feed ration does not need to be complex, only complete. Field crops will be irrigated with fish effluent during the summer in a controlled test. The control will be plain water and yields will be noted and recorded. All other parameters will be constant.

COMPOST-BASED AQUAPONIC GREENHOUSES

As vegetables are grown and sold, there will be a certain amount of waste plant parts — roots, stems, and trimmings — which will be composted along with grass clippings, leaves, and other materials the project has available, including the ashes from the outdoor furnaces. Ashes are a rich source of boron — an element lacking in most northeastern soils. The project will use the compost in a separate growing system that will allow for the growth of root crops in pure compost, and irrigated with fish tank water.

Using Vertical Space — Potatoes in Scrap Tires and Strawberries

In Towers

Potatoes and strawberries lend themselves nicely to the use of vertical space. Potatoes especially can be grown in this manner using individual stacks of waste tires to contain the growing medium (compost) and provide room for the tubers to develop. The black of the tire absorbs heat and there is usually a heavy yield. Strawberries are a more economically advantageous crop as Louisiana State University has demonstrated. Grown in vertical stacks of pots, the university was able to fit the equivalent of 10 acres of strawberries into a 6000-square-foot greenhouse. Total space for each tower is 1 square foot and the reported yield is 32 pints from each tower. The growing medium is perlite and pine bark and nutrients can be supplied through either hydroponic or fish effluent fluids.

Food that Nourishes Us

Through the use of kelp meal and solar-evaporated sea salt as ingredients in the fish feed we make, we directly add microelements to the food we are producing. The microelements remain available to the plants since there is no leaching, as in the case with soil culture. New feed each day adds more microelements to the system and thereby maintains the availability for plants and fish alike. Different vegetables have different micronutrient signatures that are made available to us when we eat these vegetables. Providing a broad range of micronutrients to the young growing plants gives each plant the opportunity to reach its full potential micronutrient signature. For instance, it is a commonly held belief that spinach has iron in it — which it does. However, it has only 10% of the iron in it than it had in 1948. As a result, you would need to eat 10 bowls of spinach today to equal the nutritional value of 1 bowl of spinach grown in 1948 (source: Internet search, Google, "bowl of spinach").

Given the above information — which I have witnessed personally — it is my belief that aquaponics — supported by green energy — will become a major provider of high-micronutrient-content food, not only in northern climates, but also in more temperate regions of the world, where wintertime heating and lighting are not a factor. In short, our breathless, visionless, juvenile, and (oh, call it not rabid) love affair with petroleum will not last forever. The resource is simply not infinite. Considering our enormous dependence on petroleum, if the last barrel of oil were sold today, what would we do tomorrow?

REFERENCES

U.S. Environmental Protection Agency, *Characterization of Municipal Solid Waste in the United States*, 1996 update, Jun. 1997.

American Petroleum Institute, http://www.api.org/edu/factsoil.htm.

S & S Aqua Farm, 8386 County Road 8820, West Plains, MO 65775, Design and Implementation Package.

U.S. Department of Energy, http://www.eren.doe.gov/biopower/faqs.html.

Gil Friend, http://www.eco-ops.com/eco-ops, 1997.

New York State Department of Agriculture and Markets, Annual Bulletins, http://www.nass.usda.gov/ny/annualbulletins.

As remembered from a newsletter published by the Taskforce for Food and Farm Policy, Albany, NY. Taskforce now defunct.

15 Conclusion

Nick L. Akers
Akermin, Incorporated, St. Louis, Missouri

CONTENTS

Abstract This chapter presents the author's opinion on the drivers and need for alternative energy sources. First, global energy use is commented on, which serves as an indicator for available resources. In-depth commentary is provided on the opportunities for fuel cells as an alternative energy source, especially as a near-term potential replacement for rechargeable batteries. Potential for fuel cells in the military and as large-scale power plants is also discussed.

INTRODUCTION

The need to fully develop alternative energy sources is apparent. The first few years of the new century have seen the highest energy prices in recent history due to geographic isolation of the primary fuel and its finite supply. Coupled with insatiable demand from emerging industrialized nations, such as China, energy has arguably become the greatest stress factor in modern society. The president of the United States held his first prime-time address in over a year in April 2005 to campaign for, among other things, the need for innovations in new sources of energy.

STATISTICAL HIGHLIGHTS

The U.S. Department of Energy's *Annual Energy Review*, published September 7, 2004, provides a wealth of statistics and research into the use, production, and availability of energy throughout the world. The "holy grail" of alternative energy

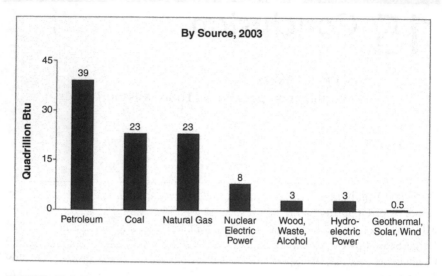

FIGURE 15.1 Energy consumption by source. *Source*: Department of Energy, Energy Information Administration: Annual Energy Review, 2003. With permission.

sources is to address the use of petroleum since it accounts for 39.4% of energy consumption followed by coal and natural gas, each accounting for 23.2% (Figure 15.1) [1].

The presidential prime-time address was not purely political; the United States is by far the largest user of petroleum, followed by Japan whose consumption is quickly being surpassed by China's budding development (Figures 15.2 and 15.3) [1].

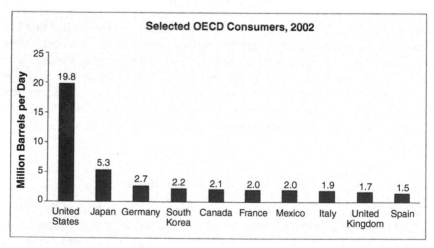

FIGURE 15.2 Petroleum consumption by country. *Source*: Department of Energy, Energy Information Administration: Annual Energy Review, 2003. With permission.

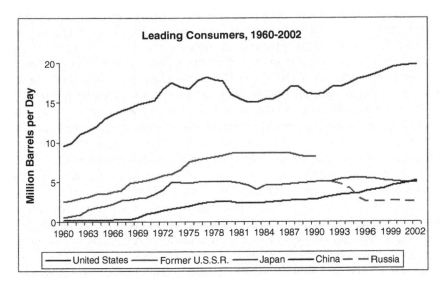

FIGURE 15.3 Petroleum consumption by country over past 40 years. *Source*: Department of Energy, Energy Information Administration: Annual Energy Review, 2003. With permission.

The United States' geographic nature of dispersed cities and urban centers has created the reliance on personal transportation and thus the leading consumption of petroleum. Despite efforts to institute mass transit, there has been little or no effect on the increasing use of this form of energy. The current administration believes the answer to be hydrogen fuel cells, stating the desire for a child born in 2003 to have a fuel-cell vehicle as their first car (January 28, 2003). However, there are nearer-term opportunities, as described in this book, which should be the focus of the country's efforts.

NEW POWER OPPORTUNITIES

There are a number of opportunities for new power sources aside from replacing petroleum for transportation. Beginning at the small end of the scale, alternatives to today's battery chemistry are in critical demand. Proliferation of portable electronics and increasing functionality has exceeded the capabilities of the lithium-ion battery. Indeed, the features able to be offered to consumers by the major electronics development companies are limited by the battery, an example being the delay of mass introduction of the 4G cell phone. Many outside of Japan may not be familiar with these phones, known as "power eaters," which last for all of 15 minutes when being used to watch television or movies. The Japanese suffer through this shortfall because their long daily work commutes are brightened by the entertainment. Batteries' limitations have at least created a new market opportunity, known as "juice bars," where the Japanese cannot only receive liquid refreshment, but for a fee can recharge their cell phones as well.

This, of course, is a temporary solution to an obvious problem: the need for new portable power sources. Fuel cells have been studied for over 40 years as a potential battery replacement technology. With a potential $4-billion-plus lithium battery market takeover opportunity, there are a plethora of entities working to deliver portable fuel cells [2]. The Direct Methanol Fuel Cell (DMFC) has probably received the majority of attention in this area. Despite hundreds of millions in research and development dollars, in early 2005 there is no consumer product on the market. DMFC has applications and viability in other markets, but the size, cost, and performance requirements for rechargeable battery replacement have thus far proved insurmountable. New catalyst developments as well as new membrane technology may push DMFCs over the hump to widespread commercialization; however, toxicity will remain a serious obstacle.

Enter the opportunity for fuels other than methanol in the portable battery market. Ethanol is the obvious choice because it is already widely available in day-to-day life and on airlines as well, a necessity. Additionally, ethanol has certain chemical properties making it desirable as a fuel, such as higher energy density than methanol while remaining a small enough molecule for good diffusion properties. As discussed in this book, innovations in catalysts are required to employ ethanol. Efficacy has been demonstrated; however, catalyst stability and operating temperature must still be addressed for metal-based systems.

Perhaps a previously considered fringe effort, now gaining momentum, is the use of nontraditional catalysts, i.e., biological catalysts. The advantages are cost savings from elimination of precious metals, simplification of system design due to high selectivity for analyte, dramatically increased fuel options, and efficient operation at room temperature, among others discussed. Enzymes in particular, have the potential to compete (and in some cases already are competing) with DMFC and DEFCs and surpass their performance. It may be breakthroughs in this area of research that ultimately enable commercial applications. Elimination of PEMs, bipolar plates, and precious metal catalysts are significant advantages.

SOFC fuel cells have also had a resurgence of effort, most likely driven by military interest. Advances in catalysts and insulating materials are showing promise for portable applications, portable meaning 20-W systems.

Overall, it is the opinion of this author that the portable fuel cell effort emerged too early for its time. Hundreds of millions of dollars, if not billions, have been taken in by start-up companies through institutional, noninstitutional, and government sources over the past 10+ years with no product to show for it. The technology continues to suffer from essentially the same key hurdles that it did since the start of effort. Many investors are losing interest and have become calloused to the excitement surrounding portable fuel cells. That is unfortunate because we are nearing the opportune time for this market. Key breakthroughs are on the verge of occurring, which will hopefully burst the bubble and pave the way for commercial applications. Some venture firms recognize this opportunity and are sticking with fuel cells in the belief that we are near the gold rush.

MILITARY

Today's military has become increasingly reliant on portable power to maintain a devastating advantage over less sophisticated enemies. Vital communications equipment, night-vision goggles, and weapon systems are being developed and deployed that require immense amounts of portable power available to the individual soldier. Lieutenant Marc Lewis was quoted in Iraq in June 2003 stating, "If we run out of batteries, this war is screwed." Soldiers are typically employing disposable batteries and some rechargeables for their equipment. Batteries can account for up to 50 pounds of a soldier's rucksack due to inability to recharge batteries in the field. To reinforce the reliance on batteries, a 12-person Special Forces team on a 30-day deployment can go through 3000 batteries at a cost of $350,000 [3]. Many of these batteries are only used for 10–20% of their capacity before being discarded. This may immediately seem wasteful but imagine staking your combative edge on being able to see at night or communicate with other troops; one would much rather pop open a new battery than use one that was not fully charged.

Portable fuel cells could provide incredible advantages to the military. Rather than carrying a number of disposable or rechargeable batteries, a solider could carry a couple fuel cells and the fuel needed to refuel them as needed in the field. Additionally, because fuel cells can provide more energy for longer periods of time than batteries, they could enable the next generation of electronic devices for the military to further enhance its combative advantage.

LARGE-SCALE POWER

Perhaps somewhat counterintuitive to the layperson, the first commercial fuel cells have been introduced for large-scale applications. Stationary power plants are being installed all across the globe. As the cost of such systems decreases and reliability increases, large-scale fuel cells will begin to be used for residential power. One fairly obvious operating concern is how to provide the fuel to residential areas. It is doubtful that such systems would operate on direct hydrogen, just as the large-scale industrial fuel cells being used today do not. Possibilities include using natural gas or piping in other liquid fuels such as methanol or ethanol using the existing infrastructure.

CONCLUSION

It would be hard for anyone to deny that energy is one of the most important issues at the start of 21st century. Energy is at the root of the major conflicts of our time as well as the catalyst for previously disadvantaged society's emergence into modern culture. As energy demand increases at staggering rates, the murmur for alternative energy technologies is quickly turning into a scream.

REFERENCES

1. U.S. Department of Energy, Energy Information Administration, *Annual Energy Review*, 2003.
2. *Darnell Group Report to the U.S. Fuel Cell Council*, Jan. 2003.
3. Valdes, J., *The World Congress on Industrial Biotechnology and Bioprocessing*, Orlando, FL, Apr. 21–23, 2004.

Index

Milton Keynes UK
Ingram Content Group UK Ltd.
UKHW020023071024
449327UK00032B/2900

9 780367 453572